Total Safety and the Productivity Challenge

Adopting a strategic approach to risk management can maximize competitiveness and profitability. Total Safety and Productivity approaches offer managers a set of methods and tools to apply a Total Safety Management (TSM) philosophy to achieve this. The capability to anticipate, assess and plan for risks associated with future operations is a critical success factor, for enterprises of all types and sizes. The ability to risk assess actual operations with an easy to apply, resilient methodology can offer significant benefits in terms of the capacity to improve safety and performance.

This book describes approaches that can be used alone or jointly to improve safety management in any organization. The methods are based on academic best practice and have been developed by leading experts, but are presented here in a practical way for application in industry by non-experts. The book outlines a professional approach to risk and safety management, which requires goal setting, planning and the measurement of performance, and encourages a safety management system that is woven holistically into the fabric of an organization so that it becomes part of the culture, the way people do their jobs, and helps ensure that issues are correctly prioritized and managed as they emerge.

This book is essential reading for professionals, at both expert and non-expert level, who are interested in applying the TSM philosophy within their organization.

Maria Chiara Leva is a lecturer in Human Factors and Industrial Safety at the Technological University Dublin. She is a visiting Research Fellow in the Centre for Innovative Human Systems at Trinity College Dublin, and is also the co-founder of a campus company called Tosca Solutions.

Tom Kontogiannis is a professor in Human Factors and Industrial Safety at the Technical University of Crete. Since 1997, he has headed the Cognitive Ergonomics and Industrial Safety (CEIS) Lab in the Technical University of Crete.

Marko Gerbec is an application research counsellor in the Department for Inorganic Chemistry and Technology, as well as an associate professor at Jozef Stefan's International Postgraduate School.

Olga Aneziris is a head of research at the Laboratory of Systems Reliability and Industrial Safety, at the National Centre for Scientific Research "DEMOKRITOS", and the project manager of various EU funded projects (e.g. INTEGRISK, TOSCA, SUPER-LNG).

Total Safety and the Productivity Challenge

Edited by Maria Chiara Leva, Tom Kontogiannis, Marko Gerbec and Olga Aneziris

LONDON AND NEW YORK

First published 2019 by Routledge
2 Park Square, Milton Park, Abingdon, Oxon OX14 4RN
52 Vanderbilt Avenue, New York, NY 10017

First issued in paperback 2020

Routledge is an imprint of the Taylor & Francis Group, an informa business

British Library Cataloguing-in-Publication Data
A catalogue record for this book is available from the British Library

Library of Congress Cataloging-in-Publication Data
Names: Leva, Maria Chiara, editor. | Kontogiannis, Tom, editor. | Gerbec, Marko, editor.
Title: Total safety and the productivity challenge / edited by Maria C. Leva, Tom Kontogiannis, Marko Gerbec and Olga Aneziris.
Description: First Edition. | New York : Routledge, 2019. | Includes bibliographical references and index.
Identifiers: LCCN 2018052668 | ISBN 9781138091306 (hardback) | ISBN 9781315108100 (ebook)
Subjects: LCSH: Organizational change—Management. | Industrial safety—Management.
Classification: LCC HD58.8 .T678 2019 | DDC 658.4/08—dc23
LC record available at https://lccn.loc.gov/2018052668

ISBN 13: 978-0-367-67175-4 (pbk)
ISBN 13: 978-1-138-09130-6 (hbk)

Typeset in Optima
by Apex CoVantage, LLC

This book is dedicated to the memory of Lorenzo Accardo

Contents

Foreword

Introduction

Over the years, safety management systems (SMS) have changed their perspectives as a result of increasing complexity of industrial plants, changing societal pressures and new theoretical risk approaches. Gradually, industrial organizations assumed a greater responsibility for their safety-critical activities that went beyond adherence to prescriptive regulation and legislation as set out by governmental inspectorates. New safety approaches have emerged that emphasized a philosophy of 'self-regulation' where safety was seen as the primary responsibility of industrial facilities. This gave rise to a promulgation of formal safety management systems and international standards of risk management that prescribed formal systems for regulating safety.

Although there are different views in the literature, there is a convergence on the following problems in putting SMS in practice in the industry.

- Safety programs often lose momentum as they take too long to demonstrate tangible benefits; this may imply that safety has not been properly integrated with other business processes (e.g., human resources, supply chain, quality).
- Safety knowledge is scattered across the organizational units, which makes it difficult to get access to risk information required for risk assessment. Hence, organizations may need to set up a suitable Knowledge Management System that makes available risk information and facilitates data transfer between risk analysis methods.
- Safety has been perceived by people as a set of administrative controls taken by safety managers and enforced by compliance with regulatory

  bodies. Hence, people may fail to see how they can contribute to safety programs or may fail to report on the effectiveness of programs.
- Plant management is focusing mostly on reactive or lagging measures of safety performance. As a result, failures in managing plant changes and safety interventions cannot be captured at early stages as they are identified until after some damage has occurred.

Even the certification of an organization with a SMS standard does not guarantee that problems in setting, running and evaluating the system will be overcome. Although surveys and benchmarks have provided some evidence of a positive relationship between SMS certification and safety performance, the evidence is not sound and it has not been confirmed by scientific studies. To ensure a positive impact of safety management systems, organizations need a set of safety capabilities that would allow them to overcome these problems.

To respond to these concerns, an effort has been undertaken by TOSCA (Total Operations Management for Safety Critical Activities), a European Project within the context of the 7th Framework Program. The scope of TOSCA was to establish an economically suitable framework in which the most innovative tools and techniques (e.g. advanced 3D software, virtual reality, innovative theoretical models, updated information exchange protocols, etc.) are used together in order to take advantage of possible synergies in processing human factors requirements, fulfilling regulations, improving safety and enhancing productivity. To achieve this, the TOSCA project developed a theoretical framework for Total Safety Management (TSM) that focuses on the principles, safety processes and techniques for promoting and integrating safety with other business processes for quality and productivity.

The Total Safety Management framework has been based on earlier efforts that include the Integral Health System (Zwetsloot & van Scheppingen 2007), the Strategic Safety Management (Zou & Sunindijo 2015) and others. A report by the Centre of Chemical Process Safety (CCPS 2006) has also provided evidence for a business case for safety. Other international risk management standards (e.g., ISO 31000) have also implied that safety processes should be integrated with other business processes to enhance organizational effectiveness. These developments have led to the concept of 'Total Safety Management', where safety, quality and productivity are addressed in a joint way during the life-cycle of a project or a product.

This book aims to present several safety and risk management processes and methodologies for the practical integration of industrial operations into a total performance management system, so that issues concerning safety, quality and productivity are optimized together.

A Total Safety Management system should be built around a set of fundamental principles and safety processes that can be distilled from three sources: (i) the requirements of an application domain (e.g., major hazards industries with emphasis on SMEs), (ii) similar approaches that have been proposed in the literature and (iii) field experience and application of the basic principles to a wide variety of case studies. The first chapter draws upon these sources of information to propose five fundamental principles of TSM to overcome different sorts of problems encountered in the implementation of earlier safety management systems. Examples include: risk perception and safety knowledge scattered around the organization, transferring risks from one unit to another, losing momentum when tangible benefits are late and not learning from the past. The five TSM principles also guide the application of a set of safety processes that have been required as part of international standards of risk management (e.g., ISO 31000). For instance, 'commitment in action' requires an integration of safety with other business processes whilst 'managing risks' requires a knowledge resource capability deployed in a participative manner. The TSM principles are applied to four broad safety processes: commitment in action, understanding risks, managing risks and learning from experience.

In some respect, a Safety Management System is a control system that adjusts its performance to achieve its objectives within certain safety constrains (Wahlstrom & Rollenhagen 2014). An important aspect of this control metaphor is that the design and application of control loops is affected by the model of safety that an organization maintains. The model of safety refers to the understanding of risks and possible mitigation measures shared among managers, supervisors and operators. Sometimes, the term Common Operational Picture (COP) is also used to refer to the mental model of the organization about what is safe or not, how weaknesses and problems can be managed, whether resources and tools are adequate do this and how things can get back to normal. Hence, Chapter 2 considers a number of challenges in implementing the COP concept such as, who should have access to the COP system, how much information to display to different users and what media should be used to portray risk information. These issues are examined in this chapter with examples from the chemical industry and then expanded further in Part 2. The following three chapters focus on 'Understanding hazards and risk' to propose to the readers an array of well-established risk assessment methods. In this sense, Chapter 3 presents a review of traditional and modern methods of risk assessment followed by a thorough description of a bowtie methodology that has been particularly effective for Small & Medium size Enterprises (Chapter 4). Finally, a new methodology is proposed in Chapter 5 for looking into the interactions between safety and other business processes.

Chapters 6–8 present the topic of how concretely a company can establish and manage a common operational picture for their risk management. This section presents examples on how to integrate safety with other business processes by aggregating a variety of performance indicators. For instance, a risk registry or a knowledge system for managing risks should be built with a learning capability in mind.

The remaining Chapters 9–12 provide an elaboration of the process of 'managing risks and change in the context of business strategy' through a variety of TSM means, such as project management, knowledge management, change management and training of process control skills.

The final Chapter 13 looks at the way forward for managing all safety processes within the complexities of modern organizations.

Understanding hazards and risks

A main element of TSM is the risk assessment process that is common to many international standards and is concerned with preparing for and then conducting a risk assessment that leads to the final treatment of the risk. Risk assessment is important because it helps create awareness of hazards and risks related to critical work activities. It purports to reduce the probability and severity of the consequences of hazards by designing suitable safety measures and precautions. Risk assessment also prioritizes hazards and helps determine how to ensure loss prevention and business continuity. Communication of risk knowledge is necessary in order for people to take responsibility for consistent implementation of control measures.

Part 2 of the book presents a synopsis of well-established methods of hazard identification and risk quantification followed by a presentation of modern dynamic risk assessment and system dynamics methods. Chapter 3 presents an overview of existing methods of hazard identification (e.g., Hazard and Operability Analysis, Failure Modes Effects and Criticality Analysis) and risk quantification (e.g., fault trees and event trees) in the context of process safety. This overview provides a basis for discussing other modern approaches to hazard identification (e.g., Recursive Operability Analysis) and risk quantification (e.g., dynamic risk assessment). Chapter 4 deals with the topic of personnel or occupational safety and presents an Occupational Risk Model (ORM) based on the bowtie methodology. Bowties have received wide recognition in both process safety and personnel safety for creating models of causes, hazards and consequences. Chapter 4 presents a well-established ORM system for modelling hazards and quantifying risks in the context of occupational safety. The bowtie models and their quantification are based on the observed historical accident data in the Netherlands; these include the accident reports of the Dutch Labour Inspectorate in their management database GISAI (2005).

It is foreseeable that interventions in safety may adversely affect other organizational interests in production, finance and quality. In other cases, safety investments may take a long time to produce tangible benefits, which could result in losing managerial enthusiasm. Understanding and modeling how changes in safety management affect the organization as a whole could provide a good basis for making better decisions about safety investments. In this context, Chapter 5 explores the use of systems dynamics in the context of TSM in order to examine the interactions of safety with other organizational domains as well as look into the time dynamics of Key Performance Indicators (KPIs) in the near future. The TSM approach to safety provides a good basis for looking deeper into the interaction of technical, workplace and organizational factors, their magnitude of effects and their time characteristics. It focuses on the organizational processes that control different goals (e.g., safety, human resources, production) and the trade-offs that should be managed in the context of the overall system. A case study of a typical coal mining operation has been examined in order to explore the feasibility of the proposed method rather than perform a realistic numerical estimation of all risks involved in mining operations.

A framework for establishing and monitoring a Common Operational Picture

Integrating safety and other business processes has been one of the main tenets of Total Safety Management, and this requires a means of measuring performance in business processes. Key Performance Indicators (KPIs) have been widely used in measuring performance and in producing a 'health check' of the organizational processes. Hence, the challenge here would be to develop risk registers and select appropriate Safety Performance Indicators that could be integrated with KPIs for other processes. Risk registers provide a means of prioritizing business goals for productivity, quality and safety as well as a means of learning from the past and achieving further improvements. In the context of this integration, Part 3 addresses many challenges in managing safety goals, generating risk profiles, selecting safety indicators across multiple industrial units, and aggregating SPIs to develop general Health, Safety and Environment (HSE) indices.

In the TSM approach, the concept of Common Operational Picture (COP) is used in Chapter 6 to structure risk information at many horizontal and vertical organizational levels, risk transfer and side effects between units and decide on the responsibility and authority levels of different stakeholders. In particular, Chapter 6 adopts a participative process of building a corporate risk register that presents safety goals, critical hazards, and measures for reducing or mitigating risk profiles. The main emphasis has been placed on how to handle streams of risk information from a large number of units, how to develop data structures that facilitate the transfer of information and how to maintain risk registers. Other challenges are also raised regarding how to decide who should participate in this process and how much information should be made available to different stakeholders. The utilized methods of data collection, risk calibration, risk assessment and information presentation are quite generic and can be applied to a wide range of industries.

The particular methodologies and mathematics for selecting, comparing, and aggregating KPIs into higher-order performance indices are presented in Chapters 7 and 8. Heidari et al. present a quantitative aggregation method based on multi-criteria decision theory that takes into account both lagging and leading indicators in the monitoring of occupational health, occupational safety, process safety, environment and HSE management. Leading indicators are particularly useful in cases where the number of hazards and incidents are extremely low. The initial set of KPIs can be selected from a baseline risk assessment that is very helpful for guiding analysis in the selection of indicators and criteria. The proposed method relies on the assessment of both negative and positive performance scores of different KPIs. When a KPI score deviates from minimum accepted level (MAL) there is a punishment procedure that takes into account the frequency and severity of deviation. On the contrary, when the performance score of KPIs are greater than a MAL score, a reward factor is calculated to compensate. In general, separate KPIs can be portrayed either individually on a radar diagram for making comparisons or collectively into a single HSE index, using the proposed aggregation method. This methodology has been applied to three gas and oil drilling companies while HSE performance trends were produced in the course of five years so as to monitor likely improvements.

Chapter 8 presents another method that is also based on multi-criteria decision theory to develop a performance management system in the design of safe, secure, cost efficient and greener inland terminals for containers with dangerous goods (ITDGS). Molero et al. have applied the Analytical Hierarchy Process (AHP) method in order to group different criteria at many levels, such as, equipment performance, business intelligence, information and communication technology, safety, security and environmental care. Instead of producing a precise decision, the AHP helps analysts find the option that best fits their objective and their knowledge. The results of the study showed that safety, security, efficiency and environmental issues were more critical for ITDGs followed by issues related to the cost of implementation and licensing of business intelligence and information technologies.

Managing projects, knowledge, skills and changes in the context of business strategy

In complex systems, safety analysts are confronted with the challenge of putting together a large number of pieces of risk information regarding the design and use of a diverse range of process systems. This integration problem exacerbates as people have their own individual ways of perception of risks, work practices and risk tolerance criteria. To overcome these issues, the TSM approach advocates the application of the principles of a Common Operational Picture (COP) and a Knowledge Management System (KMS) that support the sharing of work practices, the organization of safety knowledge and the retrieval of all documents related to safety critical tasks. Part 4 includes four chapters that deal with the particular ways of managing risks through knowledge organisation, skill acquisition, project management and change management.

Chapter 9 presents a Computerized Barrier Management System (CBMS) for managing risks at a pesticides producing plant. Barriers are safety measures for preventing or mitigating risks, including work procedures, information systems, safety devices and training. Its main objective is to demonstrate how to build a Knowledge Management System to create an electronic database of pieces of knowledge (e.g., equipment design, plant layout, manuals, permits and risk assessment procedures) that allows safety analysts to have fast access to all data sources required in risk management. The safety analysts can work on this database by applying well-established methods of risk analysis (see Chapters 3 and 4) and simple methods of workplace prototyping (e.g., 3D pictures of workstations) to manage risks. Hence, several methods of task analysis and risk assessment can be supported with the use of 3D pictures and simple virtual reality systems – e.g., preview rare procedures, identify hazards in virtual workstations and link safety barriers to 3D maps of plant sites. In this way, many parts of the risk assessment process are visualized and become accessible to non-specialists who need to contribute some knowledge to this process.

Chapter 10 makes an attempt to exploit previous uses of risk assessment methods in the domain of operator training for process control skills. Its objective has been to develop a virtual reality training environment that would be integrated with previous work produced in the context of risk assessment. Chapter 10 proposes a design framework of virtual reality training that examines how to screen safety-critical tasks for additional training, how to develop a virtual reality instructional environment, how to deliver training and how to assess the mastery of process control skills. The use of virtual reality in industrial training offers a range of benefits including, 'learning-by-doing' (when there are constraints in terms of machine availability, safety, or time) and 'cuing' with appropriate visual, auditory, or haptic cues; also, it allows simulating the task in a flexible way to adapt it to users' needs and training goals. In summary, Chapters 9 and 10 demonstrate an important thesis of the TSM approach that risk assessment methods and tools should not be developed in isolation because this could increase the cost of use and the cost of transfer of data.

To a certain extent, safety-critical activities can also be seen as 'projects' or 'safety cases' that can be considered from the perspectives of many stakeholders (e.g., different departments, subcontractors, regulatory authorities, etc.) and decision makers at different organizational levels (e.g., top managers, supervisors and operators). In this sense, the principles of TSM can also be applied to the domains of project planning and management of changes.

Chapter 11 presents a participative risk assessment process for project planning that puts in practice some established methods of task analysis and hazard identification. Project planning can be based on a range of representations ranging from Gantt and PERT charts to discrete-event simulations. To demonstrate the TSM approach to project planning, this methodology has been applied into a maintenance planning procedure. The results from the case studies indicated that significant benefits in terms of time, cost and safety can be achieved through the application of the integrated methodology.

Plant changes and organizational changes have become very common in process industries due to the increasing competition and diversification of products. The Management of Change (MOC) has become part of safety management because changes and modifications are often complex and involve interaction of technical and organizational factors. Although the Management of Change process has been well documented by several regulatory and advisory authorities, the TSM principles offer additional benefits in recognizing the need for an MOC process, planning and implementing changes as well as monitoring the results and making finer improvements. Chapter 12 applies the principles of TSM and business process design to develop a method of designing and managing change proposals that integrates the technical and organizational. The method was subject to testing and validation by two industrial organizations via ex-post re-assessments of three recent changes.

Conclusion

This book presents a systematic approach to Total Safety Management where new approaches to safety and risk management are addressing not only the negative aspects of safety (e.g., hazard identification and risk reduction) but also emphasize the positive business values that safety can bring to the organization. Although several theoretical perspectives, authority proposals and international standards of risk management have pointed out some of the principles and safety processes of TSM (e.g., 'business case for safety' and 'participative risk assessment') there is still scope for more studies to elaborate on the principles and propose new methods for applying the TSM safety processes. This has been achieved, to a large extent, with the studies presented in this special issue. It is expected that the TSM approach will inspire further studies that should make it more elaborate and applied in the context of high risk industries.

References

CCPS–The Business Case for Process Safety 2006, *American institute of chemical engineers*, Wiley & Sons, New York.

Wahlstrom, B & Rollenhagen, C 2014, 'Safety management: A multi-level control problem', *Safety Science*, vol. 69, pp. 3–17.

Zou, PXW & Sunindijo, RY 2015, *Strategic safety management in construction and engineering*, Wiley & Sons, New York.

Zwetsloot, G & van Scheppingen, A 2007, 'Towards a strategic business case for health management', in U Johansson, G Ahonen & R Roslander (eds.), *Work health and management control* (pp. 183–21), Thomson Fakta, Stockholm.

Contributors

Payam Amir-Heidari (Department of Civil Engineering, Sharif University of Technology, Tehran, Iran)

Olga Aneziris (National Centre for Scientific Research "DEMOKRITOS", Greece)

Pablo Aragonés-Beltrán and Juan-Pascual Pastor-Ferrando (Universitat Politècnica de València. Departamento de Proyectos de Ingeniería, Valencia, Spain)

Gabriele Baldissone (Politecnico di Torino, Dept. of Applied Science and Technology, Torino, Italy)

Nora Balfe (Centre for Innovative Human Systems, Trinity College Dublin, Ireland)

Dimitris Boukas (Technical University of Crete, Chania, Crete, Greece)

Maria Chiara Leva (School of Food Science and Environmental Health, Dublin Institute of Technology, Ireland)

Costanza Ciarapica-Alunni (Centre for Innovative Human Systems, Trinity College Dublin, Ireland)

Micaela Demichela (Politecnico di Torino, Dept. of Applied Science and Technology, Torino, Italy)

Marko Gerbec (Jozef Stefan Institute, Department for Inorganic Chemistry and Technology, Ljubljana, Slovenia; Jožef Stefan International Postgraduate School, Ljubljana, Slovenia)

Myrto Konstantinidou (National Center for Scientific Research "DEMOKRITOS", Greece)

Tom Kontogiannis (Technical University of Crete, Chania, Crete, Greece)

David Levovnik (Jozef Stefan Institute, Department for Inorganic Chemistry and Technology, Ljubljana, Slovenia; Jožef Stefan International Postgraduate School, Ljubljana, Slovenia)

Christina Malaki (Department of Business Administration, University of Macedonia, Salonica, Greece)

George Mavridis (K&N EFTHYMIADIS SA, Sindos, Greece)

Gemma Dolores Molero (AITEC, Research & Innovation Projects Department, Valencia, Spain)

Zoe Nivolianitou (National Centre for Scientific Research "DEMOKRITOS", Greece)

Stephen O'Mahony (ESB, Dublin, Ireland)

Ioannis Papazoglou (National Centre for Scientific Research "DEMOKRITOS", Greece)

Emmanuel Plot (INERIS, France)

Steve Prast (EOS, Solution LLC U.S.)

Francisco Enrique Santarremigia (AITEC, Research & Innovation Projects Department, Valencia, Spain)

Mehdi Sbaouni (Reviatech SAS, Parc Technologique des Rives de l'Oise, Venette, France)

Richard Sheehan (ESB, Dublin, Ireland)

PART

1

A framework of managing total safety

Total Safety Management

Why?

Tom Kontogiannis

The evolution of safe systems of work in major hazards industries

The increasing complexity of technological systems has brought about a shift in the way that safety management has been perceived in the last 40 years. In the early 1970s, worker safety at industrial plants was generally managed through an adherence to prescriptive regulation and legislation as set out by governmental inspectorates. This so-called boots, belts and buckles approach to safety management suggested that as long as organizations followed regulations with respect to the protection of the workforce, the health and safety of workers was being managed effectively. While this approach seemed sufficient at the time, a number of catastrophic accidents resulting from the complexity of the industry shifted the responsibility for health and safety from government bodies to the particular organizations and industries.

An organization's own responsibility for safety management (beyond adherence to regulatory requirements) was perhaps first proposed by the Health and Safety at Work Act in the UK (Hale & Hovden 1998). This Act set out a philosophy of 'self-regulation' where safety was seen as the primary responsibility of industry to ensure that all reasonably practical steps have been taken to improve health and safety. Although the terminology of 'self-regulation' remains problematic, this act was innovative in the sense that it focused on organizations taking responsibility for their own management of their unique risk profiles.

This shift in safety management was driven by a series of catastrophic events in a diverse set of industrial domains. For instance, the Seveso disaster (1976) gave rise to the Seveso II directive which mandated systematic management systems across facilities in Europe that handled dangerous substances. A decade later, the Piper Alpha accident provided a context for putting in place similar directives for offshore oil and gas facilities as a result of the Cullen inquiry. In this way, responsibility for health and safety became the primary concern of organizations, with regulatory bodies becoming the evaluators of organizational efforts in safety. Only when the primary mechanism for safety management moved from prescriptive regulation to organizational responsibility were specific safety practices grouped together, under the banner of Safety Management Systems (SMS). In this sense, modern SMS could be defined as collections of best practices and strategies by which organizations demonstrate that they are taking all reasonably practical steps to ensure the safety and welfare of their employees.

In the 1990s, we have witnessed an increase of national standards trying to support industries in their efforts to control and improve health and safety performance. An example of a widely used system has been the Successful Health and Safety Management System (HSG65) which was first prepared by the Health and Safety Executive (UK) as a practical guide for directors, managers, health and safety professionals and employee representatives who wanted to improve health and safety in their organizations. Because many national efforts remained largely uncoordinated, an international collaboration called the Occupational Health and Safety Assessment Series (OHSAS) Project Group was formed to create a unified approach. Drawing on the best of existing standards and schemes, the OHSAS Project Group published the OHSAS-18000 Series and received wide recognition in the industry.

In fact, many national standards and the OHSAS-18000 series mirrored several components of the ISO 9001 series pertaining to quality management systems and the ISO 14001 series pertaining to environmental management systems. Indeed, for some time organizations have sought an integration of their management systems in order to increase efficiency and avoid duplication of processes across the organization.

Although, several 'safety programs' have been advocated in the period 1990–2000 that looked into separate aspects of safety management (e.g., safe work practices, mechanical integrity, hazards identification, incident analysis) the modern Safety Management Systems that emerged in the 21st century provided a more comprehensive framework that puts together all these safety programs. The SMS provided industrial organizations with a system that facilitated the transfer of information between different programs, defined an organizational structure of people responsible for the programs and optimized allocation of resources for programs and integrated them in ways that maximized effectiveness. It should be noted, however, that SMS provide only a structure for putting together various safety programs but do not propose specific methodologies, tools and technologies for managing the system.

There have been a few publications about the sort of problems encountered by major hazards industries in implementing safety management systems (Hardy 2010; Roderick 2006). Although there are different views in the literature, there is a convergence on the following problems in putting SMS in practice in the industry.

- Safety programs often lose momentum as they take too long to demonstrate tangible benefits and high level management gets absorbed by other production requirements that are seen at odds with safety. This is mainly due to the fact that safety is not usually integrated with other business processes and appears to be an extra workload for the organization.
- Safety knowledge is scattered across the organization with people holding different views on what constitutes a significant hazard, how plant changes and risk measures may have local benefits but produce side effects at other units, who is responsible for managing safety at different units, what safety competences exist on other units and so forth.
- Safety has been perceived by people as a set of administrative controls taken by safety managers and enforced by compliance with safety procedures. Hence, sharp-end operators may fail to see how they can contribute to safety programs or may fail to report on the effectiveness of programs.
- Difficulties have been reported at choosing the right methods for hazard identification and risk analysis as well as finding appropriate data from documents, work permits, procedures and plant modifications. To overcome these problems, organizations need to set up a suitable Knowledge Management System (KMS) that makes available risk information and facilitates data transfer between risk analysis tools.
- Monitoring the effectiveness of safety programs and assessing the safety performance of organizations remain important challenges, with most organizations focusing only on reactive or lagging indicators. As a result, failures in managing plant changes and safety interventions cannot be captured at early stages as they are identified after some damage has occurred.

Even the certification of an organization with an SMS standard does not guarantee that problems in setting, running and evaluating the system will be overcome nor that there is going to be a continuing improvement in safety performance. For this reason, organizations with their own methods and ways of implementing OHSAS-18000 have achieved different results. Although surveys and benchmarks have provided some evidence of a positive relationship between SMS certification and safety performance, the evidence is not sound and has not been confirmed by scientific studies. To ensure a positive impact of safety management systems, organizations need a set of safety capabilities that would allow them to overcome the above problems.

This scientific effort has been undertaken by TOSCA (Total Operations Management for Safety Critical Activities), a European Project within the context of the 7th Framework Program aimed at developing the sort of safety capabilities that would enable organizations to make their SMS work in practice. The scope of TOSCA was to establish an economically suitable framework in which the most innovative tools and techniques (e.g., advanced 3D software, virtual reality, innovative theoretical models, updated information exchange protocols, etc.) are used together in order to take advantage of possible synergies in processing human factors requirements, fulfilling regulations, improving safety and enhancing productivity.

To achieve this, the TOSCA project developed a theoretical framework for Total Safety Management in the process industry, with a particular focus on Small and Medium sized Enterprises (SMEs). Total Safety Management (TSM) focuses on the principles, safety processes and techniques for promoting and integrating safety with other business processes for quality and productivity. This chapter looks at the guiding principles and safety processes for managing 'total safety' while appropriate techniques and case studies follow in the remaining chapters.

Proponents of Total Safety Management

Many Safety Management Systems currently in use by major hazards industries have been developed on the basis of ISO standards, such as OHSAS-18000, hence adopting a more 'self-regulating' style where safety becomes the primary responsibility of organizations. In recent years, we have witnessed new frameworks of safety management that provide a useful basis for developing a 'performance-oriented' approach that integrates safety with other business processes, sets goals for safety, defines the context of risk assessment and evaluates safety intervention with respect to safety goals. To this end, this section reviews a few modern approaches to safety management that have implications for building a Total Safety Management approach.

For many years, the Total Quality Management (TQM) principles have provided a basis for developing several health and safety systems. Building on TQM, Goetsch (1998) introduced the concept of Total Safety Management (TSM) as a performance-oriented approach that gives organizations a sustainable advantage in the marketplace by establishing a safe work environment that is conducive to peak performance and continual improvement. The fundamental elements of TSM include: strategic approach to safety, emphasis on performance assessment, employee empowerment, reliance upon robust methods of risk analysis and continual improvement. However, more specific and practical organizational processes for total safety have been proposed by the Strategic Safety Management (SSM) approach that emphasized an integration of safety into the corporate strategy and a demonstration of business values of safety (Rahimi 1995; Zou & Sunindijo 2015).

In a total safety approach, business processes are integrated with safety engineering techniques within a continuous improvement culture that affects all levels in the organization. In the SSM approach, the safety target becomes the analysis of 'work processes' rather than the analysis of isolated safety critical activities. A 'work process' is a complex web of interdependencies between physical entities, information, communication and knowledge channels and decision-making activities. Hence, by analyzing what is wrong with a work process, safety practitioners can evaluate the entire system and cater for safety, quality and productivity. To exploit operational feedback, the SSM relies on performance measures that relate to work processes rather than work outputs (e.g., incident and injury rates). These SSM principles have been applied by Zou and Sunindijo (2015) in the development of safety programs for the construction industry.

In the Integral Health Monitoring system (IHM), the value of health is seen as a key element of corporate policy in addition to the reduction of incidents and their associated costs (Zwetsloot 2003; Zwetsloot & van Scheppingen 2007). The IHM principles have been based on earlier TQM approaches and Business Excellence Models which provide a good basis for integrating safety with quality and other business processes. The IHM approach requires a shift from solving safety problems and reducing risks to the positive business values that safety can bring to the organization. The focus is no longer on risk reduction, medical problems or product safety but on a combination of them and their relationships to organizational and business development. In this sense, health and safety is associated with business values that increase its strategic role.

Recently, there has been an effort to integrate existing frameworks for quality (ISO 9001), management of safety (OHSAS-18000), and environmental management (ISO 14001) within a single Integrated Management System (IMS). Reasons for implementing an IMS approach include eliminating conflicts, avoiding redundant procedures and enabling a synergy among standards (Bernardo et al. 2015). The ISO 31000:2009 standard for risk assessment serves to unite different risk management processes with existing standards of quality and environmental management and offers a common approach to address risks, without necessarily leading to a process of certification. The ISO 31000 standard can be applied throughout the life of an organization for a wide range of activities including strategies and decisions, operations, processes, functions, projects, products, services and assets. In particular, the principles of effective risk management in ISO 31000 have been summarized in Table 1.1.

The application of ISO 31000 to SMEs is an important topic because SMEs have a high economic potential for European countries as they provide

Table 1.1 Principles of effective risk management according to ISO 31000

1.	Risk management (RM) should create and protect business values
2.	RM should be central to the organization's processes
3.	All decision-making within the organization involves the explicit consideration of risks and the application of risk management to some appropriate degree
4.	RM should be based on best available information
5.	Continual communication with external and internal stakeholders, including comprehensive reporting of safety performance
6.	RM should be comprehensive and clear about accountability for risks, controls and risk treatment processes
7.	It should be systematic, structured and timely applied to critical activities
8.	It should take into account human and cultural factors
9.	It must be dynamic, iterative and responsive to change
10.	It must facilitate continual improvement of the organization

many opportunities for employment and economic growth. There are many concerns in performing risk assessment activities because of limited financial resources, lack of appropriate competence and management tools and a burden of compliance with regulations and codes. Therefore SMEs can improve safety performance when they develop a capability to assess risks in actual operations and a capability to monitor daily risks.

Principles of Total Safety Management

A Total Safety Management system should be built around a set of fundamental principles that can be distilled from three sources: (i) the requirements of an application domain (e.g., major hazards industries), (ii) similar approaches that have been proposed in the literature and (iii) field experience and application of the basic principles to a wide variety of case studies. In this respect, the last source of information was a three-year effort of applying the TSM approach to several case studies in the TOSCA project.

The five earlier concerns about existing SMS approaches implicate five fundamental principles for TSM that map into the ten ISO 31000 principles (Table 1.1) as follows:

1 TSM should be part of all decision-making and organizational processes and provide a capability for creating a business case for safety.
(#1,2,3 Table 1.1)

2 TSM should be based on best available risk information to create a common operational picture about risks.
(#4)

3 Participative risk management must ensure that all stakeholder needs are considered while their knowledge about risks is brought into play.
(#5)

4 Knowledge management should be part of TSM so that all knowledge about risks is managed effectively and all risk analysis techniques are better integrated.
(# 6,7,8)

5 Performance monitoring and operational feedback is necessary for making TSM dynamic, iterative and responsive to change. At the same time, this will facilitate continual improvement of the organization.

(# 9,10)

The five TSM principles are further elaborated below, using existing knowledge from risk management standards, earlier approaches to total safety and our recent experience with a three-year European project (TOSCA) that applied the TSM principles to develop useful processes and methods in safety management.

Business case for safety

For many years, several institutions and regulatory authorities have emphasized the need for safety management to demonstrate a business value to organizations. For instance, the Health & Safety Executive (HSE, UK) provided some evidence that 'Good Health Is Good Business' (HSG65 publication) since work accidents can come at a high cost to the business. In a benchmark study with the chemical processing and petroleum industries, the Centre for Chemical Process Safety (CCPS 2006) found that companies that had an efficient SMS in place achieved many returns from their investment such as productivity increases, production and maintenance cost decreases, lower capital budgets and lower insurance premiums. It appears then that we should seek to set out a much broader 'business case' for safety showing how a proactive safety approach can benefit the business by improving quality and reliability, encouraging workforce innovation and enhancing corporate reputation.

Common Operational Picture (COP)

Safety management depends upon all personnel building a common understanding of what constitutes a hazard or risk in a facility, what safety controls can minimize risks, how one's own job may produce hazards at other places and how it is possible to evaluate the current state of process activities. This safety knowledge may be scattered around the organization but it should be coordinated by means of a common understanding or picture in the organization. Some authors have used the term Common Operational Picture (COP) to refer to this 'mental model' of the system, its hazards and its safety controls. A COP may be a single source, usually a display of relevant operational information about risks. Finding what constitutes a 'common operational picture' remains a challenging issue in a safety management system. For instance, 'risk registers' (see Chapter 6) provide risk profiles and review the effectiveness of earlier safety interventions, hence facilitating a common understanding of the whole spectrum of risks in an installation. The COP may be viewed as a knowledge base of risk information that is accessible to many practitioners but 'who should access what information' remains a challenge.

Participative risk assessment

Risk assessment (RA) is a complex process performed by specialized people who are experts in engineering and reliability aspects of system design. Operational problems in the use of equipment that may produce work hazards are usually based on descriptions from operating procedures. This engineering approach, however, may miss out any informal practices with safety repercussions since these are not documented in procedures. In most cases, a risk assessment is performed on how tasks should be carried out rather than on how they are actually performed by operators. As a result, critical diversions or violations of procedures are missed in this analysis. To avoid this oversight, a participative risk assessment is required that would engage people at all hierarchical levels in certain stages of the analysis. This approach ensures that analysts capture all risks associated with how the work is done. Another advantage would be that operators are motivated to spot and report hazards, which updates and improves any previous risk assessment studies.

Knowledge management

Several authors have advocated the idea of a Knowledge Management System (KMS) with a learning module that can improve health and safety practices (Sherehiy & Karwowski 2006; Hugenholtz et al. 2007; Podgorski 2010; Floyde et al. 2013). This approach views the knowledge capital of practitioners as yet another resource that could be used to support safety management. An e-learning facility can assist companies in collecting and disseminating knowledge about safety matters across organizational levels. Additional functions of the KMS could include: knowledge about operating complex equipment, failure modes and operational hazards, job competencies for various operations and work documents (e.g., operating procedures and work permits). In most installations, this knowledge is kept in silos which hinders wide access to many practitioners. KMS can bring together different plant knowledge, reshape its format to provide better integration and organize knowledge in functional and user-friendly ways.

Performance monitoring and feedback

New safety standards (e.g., ISO 31000) have highlighted the important role of performance monitoring and operational feedback in the safety control loop. Most industries usually keep a record of safety indicators – e.g., Incident Rate, Potential Loss of Life, Fatal Accident Rate, Lost-Time Injury Rate – to determine the effectiveness of risk mitigation measures. Unfortunately, 'lagging indicators' do not provide any indication of the safety capabilities of an organization until it is too late. For this reason, companies need to specify a set of indicators that have a prognostic value – that is, they can provide a general 'health check' of the state of affairs in an industrial facility. There is a need, therefore, for 'leading indicators' that issue early warnings so that safety interventions are implemented to mitigate risks in a timely fashion. Recently, there has been a large literature on 'leading indicators' that measure safety processes rather than safety outcomes (Hopkins 2009; Oien et al. 2011; Reiman & Pietikäinen 2012). Certainly, a mixture of lagging and leading indicators can be used to provide a 'health check' on a continuous basis using several graphics that create functional groups of indicators (see Chapters 7 and 8).

TSM safety processes

The safety principles of Total Safety Management provide general guidance to develop a set of safety processes that specify, plan, implement and evaluate safety programs. The central spine of TSM is the risk assessment process that is common to many international standards and is concerned with preparing and conducting a risk assessment that leads to the final treatment of the risk. Organizations manage risks by identifying them, analyzing them and evaluating whether the risks should be reduced with special treatments in order to ensure sustainability. These safety processes have been

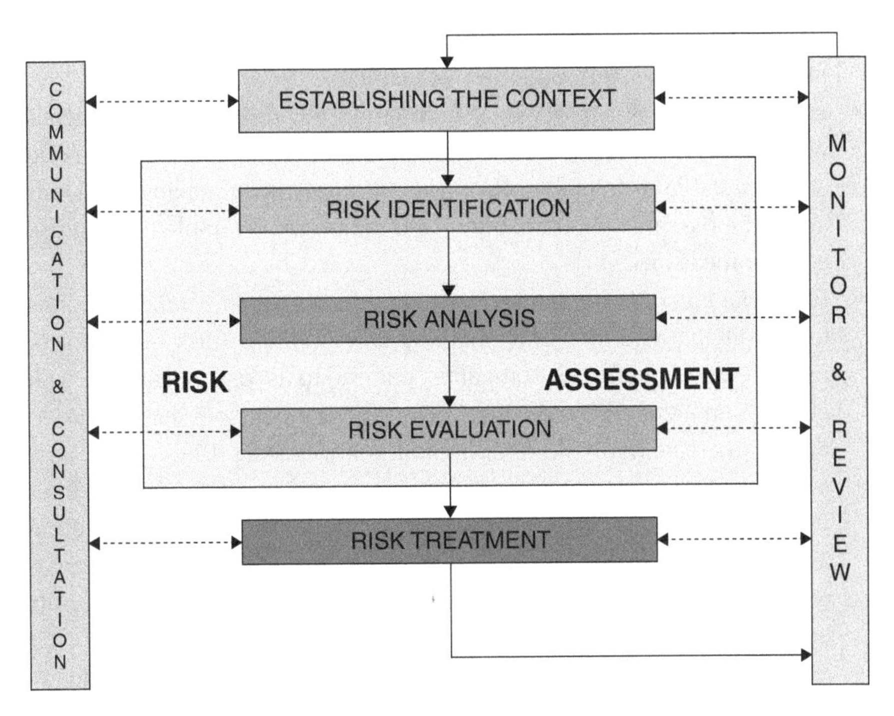

Figure 1.1 Risk management processes (ISO 31000)

expanded in ISO 31000 in order unite existing standards of quality, environmental management and safety management. Figure 1.1 provides a flow diagram of safety processes that can become the basis for a TSM approach.

A brief description and definition of key risk management processes is provided below.

- *Establishing the context* concerns the definition of external and internal parameters to be taken into account when managing risks, and setting the scope and criteria in the risk management policy.
- *Communication and consultation* is a continual and iterative processes that an organization conducts to provide, share or obtain information and to engage in dialogue with stakeholders regarding the management of risks.
- *Risk identification* involves the identification of risk sources, events, causes and potential consequences. Risk identification may involve

historical data, theoretical analysis, informed expert opinions, and stakeholders' needs.
- *Risk analysis* involves developing and understanding risks and how they impact the organization. It is expressed in terms of the consequence and likelihood of risks and relies on risk matrices.
- *Risk evaluation* is a process of comparing the results of risk analysis with risk criteria to determine whether the risk and/or its magnitude are acceptable or tolerable. Risk evaluation assists in making decisions about the risks that need treatment and the priority in implementing risk counter-measures.
- *Risk reduction or treatment* involves the selection of one or more options for modifying risks and implementing those options. Risk treatment of negative consequences is sometimes referred to as 'risk mitigation', 'risk elimination', and 'risk reduction'. It is important to note that risk treatment can create new risks in trying to modify existing risks.
- *Monitoring* involves regular checking or surveillance of risks and their proposed treatment. Example questions include: Are controls operating as they were planned? Have circumstances arisen that expose the organization to greater risk? Have lessons been learnt? Have internal audits been implemented?
- *Reviewing* is an activity undertaken to determine the suitability, adequacy and effectiveness of the subject matter to achieve established objectives.

These management processes of ISO 31000 have been used to develop the TSM processes (Figure 1.2) so that this proposal is also compatible with the four pillars of safety management advocated by CCPS (2007). The TSM processes are centred on a Common Operational Picture (COP) which links together the four areas of commitment in action, understanding risks and hazards, managing/treating risks and learning from experience.

Commitment in action

The first TSM pillar concerns the 'context of risk management', that is, the processes and organizational parameters to be considered when managing risks. For high-risk activities, organizations should also consider the depth of risk assessment, the facility areas for further analysis, the appropriate techniques and the safety investments to be made. It is also important that organizations decide on the 'risk criteria' that should determine risk tolerance levels. It is

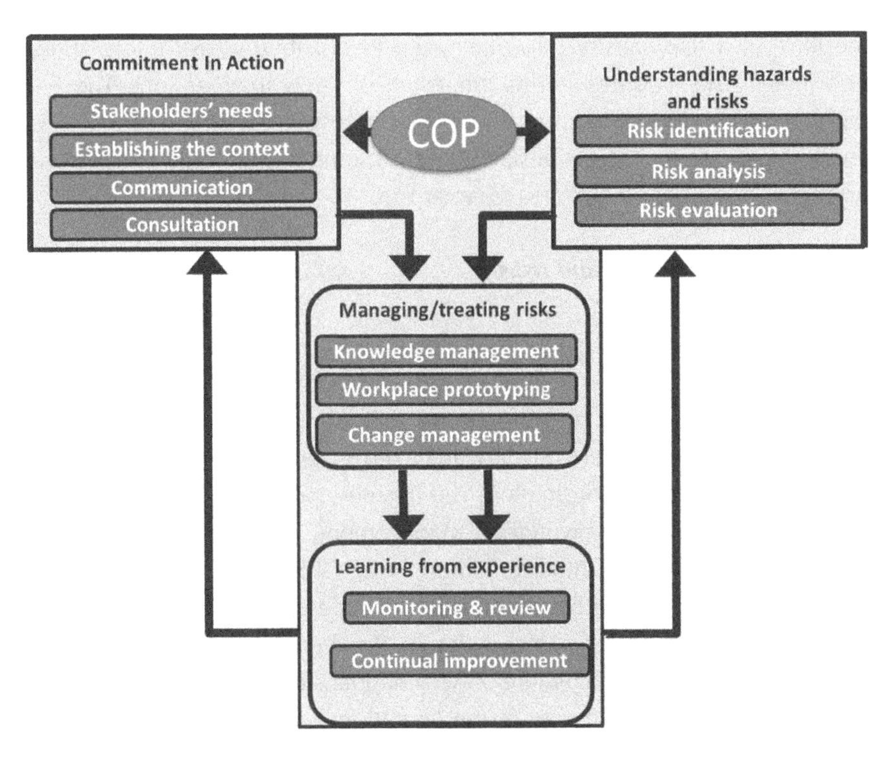

Figure 1.2 The four pillars of TSM and their enabling safety processes (adapted from CCPS 2007)

anticipated that commitment in action and risk requirements are better served when a process of communication and consultation process is in place. In order to make Safety Health and Environment (SHE) programs successful, managers and supervisors should combine health and safety with other business processes, measure safety performance, evaluate organizational changes and invite workforce participation. To close the safety control loop, practitioners at the sharp end should provide early feedback on risks in a reliable fashion.

The Common Operational Picture (COP) is very important in fostering commitment in action and in identifying processes and organizational parameters that need to be accounted for when managing risks. The COP could become the basis for developing a risk model to ensure a common understanding of hazards within the organization so that practitioners have a common mind-set. The COP or 'risk picture' shall be understandable by all relevant personnel, decision makers as well as engineering and/or operating personnel. In the TOSCA project, the COP was practically realized in the form of a risk registry for collecting and distributing risk information, prioritizing risks, and monitoring the results of safety interventions. The risk register can support a continuous learning process where risk information is regularly updated, progress of safety interventions is monitored and lessons from near misses are learnt (see Chapter 6).

Understanding hazards and risks

In TSM, risk assessment should be based on a functional system model which describes both the activities of the operators and the business processes that support human performance (i.e., supplies, tools, workplaces, training). The human operations are analyzed in terms of tasks, information needs, decisions and coordination. The business processes are analyzed in terms of task allocation, workload management, training, teamwork and team coordination. TOSCA has relied on the SCOPE tool (Supplies, Process, Context, Organizing and Effects) to drive a process model that is based on the requirements for re-design and change (Leva et al. 2015). The model of the system provides a rich picture of what happens in normal practice inside the organization. It is a model of the operational system and its processes which incorporates rich layers of knowledge about how the system works.

The functional system model provides useful information for the risk model which includes a repertoire of hazard identification, risk analysis and risk evaluation approaches as explained below:

- *Hazard Identification*. The ISO 31000 standard calls for methods/tools to identify risks under direct or indirect influences of the organizational processes. The exact methods and tools depend on the needs of the organization. These needs can be identified by a participatory approach between the risk assessors and the other stakeholders. In the TOSCA project, many case studies have used 3D modelling tools to assist the process of hazard identification. With 3D modelling tools, for instance, plans can be explored before they are implemented allowing analysts to recognize risks and hazards that were not considered with traditional risk assessments.
- *Risk Analysis*. ISO 31000 provides limited guidance on the actual process of risk assessment, requiring only that a risk assessment is carried out that fully accounts for reported hazards and their likelihood. The risk registry and the Knowledge Management Systems (KMS) may assist SMEs in conducting a practical risk analysis. The risk registry usually stores

individual risk analyses in a central repository while the KMS ensures that assessors have adequate information to make a good assessment of the level of risk within an organization.
- *Risk Evaluation.* This involves a comparison between the level of risk that has been calculated in risk analysis and the agreed criteria of risk acceptance, as decided earlier in the context of risk assessment. In a similar way, the risk registry may be useful in plotting the assessed risks onto a risk matrix, or a similar interface, allowing managers to quickly identify areas of risk that are unacceptable to the organization.

ISO 31000 requires that risk assessment is validated against operational experience (i.e., operating staff can validate the results) and external experience (i.e., other risk analysis performed in similar installations and regulatory guidelines). Another method of validation regards the use of workplace prototyping tools that provide a realistic 3D work environment where operator actions and errors can be recorded for further analysis and discussions. In the TOSCA project, virtual prototyping tools have been used for the assessment of procedures (Chapter 11), for the generation of a Computerized Barrier Management System (Chapter 9) and for operator training in a car truck unloading scenario (Chapter 10). The virtual reality workstations allow analysts to identify work hazards, monitor the performance of operators and collect human error data about specific scenarios.

Managing or treating risks

Total Safety Management considers a wider range of risk management to the risks described in ISO 31000, particularly in the area of risk mitigation. Managing risks includes three processes:

- Managing risks arising from usual variations of the performance of people and systems in everyday situations (*'ordinary risks'*)
- Managing risks arising from side effects of safety interventions and organizational changes to control or mitigate hazards (*'residual risks'*)
- Managing risks appearing in emergency situations during response management (*'emergency response risks'*)

Managing risks basically involves setting safety goals and designing barriers to prevent, control and mitigate risks. To achieve this, a risk model should be used to spot hazards, consider accident sequences, and evaluate safety barriers and distal factors that affect the likelihood of accidents. Risk models may be initially developed in the form of bowties that are widely understood by many supervisors and operators (see Chapter 4). Risk models usually require a good knowledge of technical processes, interactions and operations, reliability data and factors affecting performance. In this respect, a Knowledge Management System (KMS) can be useful to assess safety barriers, design new barriers and monitor their functioning after implementation. The KMS stores all relevant plant knowledge into a single database and provides a better organization of knowledge so that organizational changes and interventions do not produce any side effects to other parts of the system.

The competitive and financial environment in which major hazards industries operate may imply that SMEs and larger companies are constantly in a process of production changes and organizational changes to meet new requirements. Due the criticality of the 'management of change' (MOC) process, there has been a formal requirement for well-specified procedures of 'management of change' with reference to technical issues by the EU Seveso directive 2012/18/EU. Total Safety Management provides a structured approach for managing plan and organizational changes as illustrated in Chapter 12.

Learning from experience

'Learning from experience' becomes a feedback loop to the TSM system in the efforts of managers, supervisors and operators to achieve a satisfactory level of safety. Learning involves monitoring of changes or interventions and reviewing of the effectiveness of safety processes (e.g., risk assumptions and risk tolerance criteria). Although many SMEs may have understood the importance of safety monitoring, they rarely know what data to collect, what safety parameters are important, how often to collect data and how to guarantee data reliability and confidentiality. As a result, several problems may arise in monitoring safety. For instance, many workers may hold important risk information but they do not know when and how to report it. Other cases involve the reporting of near misses in standard forms without the identification of causal factors. Last but not least, raw data about safety are collected but it remains difficult to make sense of them.

The fourth TSM pillar also includes risk communication to the workforce and safety training of practitioners at the sharp end. Risk communication involves posting risk information throughout the organization and receiving feedback from the operators at the sharp end. Safety training (e.g., classroom courses or apprentice training) complements risk communication for critical situations that provides important safety skills. A more comprehensive form of training may use virtual reality representations or simulators of the workplace to allow safe practice of process control skills.

It is also important that risk assessment and training are integrated in a meaningful way so that operator skills in avoiding, recognizing and controlling hazards are identified in risk analysis and fed into a training needs analysis phase. This procedure ensures that the objectives and needs of training are tailored to the risk profiles of different jobs and that evaluation criteria for training effectiveness are derived in a systematic fashion. In this respect, Chapter 10 presents a procedure for integrating risk analysis and training in the context of critical un/loading scenarios of hazardous materials.

Concluding remarks

The TSM framework adopts a 'systems view' of safety that involves setting up safety goals, guiding risk assessment with a COP, applying formal safety and coordination rules and monitoring results through many feedback channels (Leveson 2011; Wahlstrom & Rollenhagen 2014). In this framework, setting safety goals is integrated with other business objectives while a COP supports a shared understanding of risks and safety processes in action. The principle of user participation and knowledge management also enriches the risk assessment process and grants safety analyst more choices in safety mitigation measures. Finally, the principles of consultation and monitoring/review provide further support to the other principles as advocated by ISO 31000.

People may wonder how mature the TSM framework is and whether the five principles are adequate in dealing with complex situations. In this book, our efforts have focused on coherently integrating a small number of TSM principles, using several information technology facilities for operationalizing new concepts (e.g., business case for safety, COP and knowledge management) and older ones that have been well addressed in the literature (e.g., participative risk management and performance monitoring). It is anticipated that the TSM framework can be enhanced with more principles by other approaches to the extent that these are well integrated together and associated with methods and tools for practical implementations. An example can be the principles of Complex Adaptive Systems (CAS) that can expand the systems view of safety (Reiman et al. 2015).

The following chapters present several case studies where a TSM system can be built with manageable resources. The chapters also reveal additional challenges in building TSM systems as they suggest ways of managing them. Among others, a critical challenge relates to the validation of the database as users would expect to find trustable and validated information to carry out their safety-related activities. On the one hand, the TSM system should allow the free reporting of risk information by the community of experts and users. One the other hand, this tacit knowledge of experts has to be controlled in such a way that only proofed information can be made available. Hence, there is a need to develop methods that integrate tacit and written knowledge and validate the content of the TSM system. This challenge has been undertaken in many case studies that are described in the following chapters.

References

Bernardo, M, Simon, A, Tari, JJ & Molina-Azorin, JF 2015, 'Benefits of management systems integration: A literature review', *Journal of Cleaner Production*, vol. 94, pp. 260–7.

CCPS–The Business Case for Process Safety 2006, *American institute of chemical engineers*, Wiley & Sons, New York.

CCPS–Guidelines for Risk Based Process Safety 2007, *American institute of chemical engineers*, Wiley & Sons, New York.

Floyde, A, Lawson, G, Shalloe, S, Eastgate, R & D'Cruz, M 2013, 'The design and implementation of knowledge management systems and e-learning for improved occupational health and safety in SMEs', *Safety Science*, vol. 60, pp. 69–76.

Goetsch, DL 1998, *Implementing total safety management: Safety, health and competitiveness in the global market*, Prentice Hall, Upper Saddle River, NJ.

Hale, AR & Hovden, J 1998, 'Management and culture: The third age of safety: A review of approaches to organizational aspects of safety, health and environment', in AM Feyer & A Williamson (eds.), *Occupational injury: Risk prevention and intervention*, Taylor & Francis, London.

Hardy, TL 2010, *The system safety skeptic: Lessons learned in safety management and engineering*, Author House, Bloomington.

Hopkins, A 2009, 'Thinking about process safety indicators', *Safety Science*, vol. 47, pp. 460–5.

HSG65-Health & Safety Executive 1997, *Successful health and safety management*, 2nd edn, Health & Safety Executive, Sudbury, UK.

Hugenholtz, NIR, Schreinemakers, MA, A-Tjak, MA & van Dijk, FJH 2007, 'Knowledge infrastructure needed for occupational health', *Industrial Health*, vol. 45, pp. 13–18.

ISO-9001 2008, *Quality management systems: Requirements*, The International Standards Organisation, Geneva, Switzerland.

ISO-14001 2004, *Environmental management systems: Requirements with guidance for use*, The International Standards Organisation, Geneva, Switzerland.

ISO-31000 2009, *Risk management*, The International Standards Organisation, Geneva, Switzerland.

Leva, MC, Naghdali, F, Balfe, N, Gerbec, M & De Michela, M 2015, 'Remote risk assessment: A case study using SCOPE software', *CET: Chemical Engineering Transactions*, vol. 43.

Leveson, N 2011, *Engineering a safer world: Systems thinks applied to safety*, MIT Press, Cambridge, MA.

OHSAS-18000 1999, *Occupational health and safety management systems: Specification*, British Standards Institution, London.

Oien, K, Utne, IB & Herrera, IA 2011, 'Building safety indicators: Part 1 Theoretical foundation', *Safety Science*, vol. 49, pp. 148–61.

Podgorski, D 2010, 'The use of tacit knowledge in occupational safety and health management systems', *International Journal of Occupational Safety and Ergonomics*, vol. 16, pp. 283–310.

Rahimi, M 1995, 'Merging strategic safety, health and environment into total quality management', *International Journal of Industrial Ergonomics*, vol. 16, pp. 83–94.

Reiman, T & Pietikäinen, E 2012, 'Leading indicators of system safety: Monitoring and driving the organizational safety potential', *Safety Science*, vol. 50, pp. 1993–2000.

Reiman, T, Rollenhagen, C, Pietikäinen, E & Heikkila, J 2015, 'Principles of adaptive management in complex safety-critical organizations', *Safety Science*, vol. 21, pp. 80–92.

Roderick, P 2006, 'Pitfalls in implementing and using safety management systems and behavioural modification programmes', *IChemE Symposium series no. 151*, Hazards XIX, 28–30 March.

Sherehiy, B & Karwowski, W 2006, 'Knowledge management for occupational safety, health and ergonomics', *Human Factors and Ergonomics in Manufacturing*, vol. 16, pp. 309–19.

Wahlstrom, B & Rollenhagen, C 2014, 'Safety management: A multi-level control problem', *Safety Science*, vol. 69, pp. 3–17.

Zou, PXW & Sunindijo, RY 2015, *Strategic safety management in construction and engineering*, Wiley & Sons, New York.

Zwetsloot, G 2003, 'From management systems to corporate social responsibility', *Journal of Business Ethics*, vol. 44, pp. 201–207.

Zwetsloot, G & van Scheppingen, A 2007, 'Towards a strategic business case for health management', in U Johansson, G Ahonen & R Roslander (eds.), *Work health and management control* (pp. 183–213), Thomson Fakta, Stockholm.

Understanding hazards and risks

The need for a Common Operational Picture

Tom Kontogiannis and Maria Chiara Leva

Introduction

Many major hazards industries have come to rely on Safety Management Systems (SMS) to resume responsibility for identifying, assessing and controlling their risks within acceptable levels of tolerance. The increasing demand for SMS has produced formal systems where safety targets are set out, procedures are specified for identifying and managing risks, error reduction measures are implemented and work progress is evaluated to evaluate the efficiency of safety interventions. In some respect, a Safety Management System is a control system that adjusts its performance to achieve its objectives within certain safety constraints (Wahlstrom & Rollenhagen 2012). An important aspect of this control metaphor is that the design and application of control loops is affected by the model of safety that the particular organization maintains. The model of safety refers to the understanding of risks and possible mitigation measures shared among managers, supervisors and operators. Sometimes, the term Common Operational Picture (COP) is also used to refer to the mental model of the organization about what is safe or not, how weaknesses and problems can be managed, whether resources and tools are adequate to this and how things can get back to normal.

The COP has been implemented with success in military environments for managing large-scale operations where multiple agencies have to coordinate their efforts. A COP is a representation media that can collect, process and manage information of real-time battlefield information (Liu et al. 2011). The COP concept has also been extensively applied in the context of emergency response planning and crisis management where information may be incomplete, late or unreliable. As reported by Luokkala et al. (2017), the COP incorporates information which enables situational information to be produced, visualized and presented in such a way that all information is available to all the actors involved in the crisis response in real time. A common repository and display of information is very important for command and control operations as multiple teams are trying to coordinate and respond quickly to manage a situation where uncertainty is high, available time is limited and errors can have severe consequences. In this respect, a COP system can support diverse teams in acquiring situation awareness and coordinating effectively towards a common plan.

A similar argument can be made for the role of COP in providing a shared model of 'how the system works' that guides the application of Safety Management Systems in major hazards industries. Plant personnel may find themselves in a range of situations calling for safety decisions that range from routine operations to minor equipment problems, process disturbances and emergency situations. Although these situations may vary in time pressure, familiarity and criticality, personnel always have to collect relevant information, communicate orders and requests and make a final decision how to solve the problem. In this sense, a COP system can support people in assessing the situation and applying the procedures of an SMS.

The COP has been a central concept in Total Safety Management (TSM) systems because it links together the processes of goal setting and commitment, understating risks, managing risks and learning from the past. It is important therefore to consider what sort of safety information should be collected in a COP system, who should have access to it, how much information to display to different users and what media should be used to portray risk information. These issues are examined in this chapter with examples from the chemical industry whilst another case study is presented in Chapter 9.

The following section considers the role of the COP as a model for anticipating risks and providing feedback on corrective actions. Next, earlier applications of the COP concept in emergency response planning are presented where the incident command team can make better sense of the information scattered in different places. The final section looks at different uses of the COP concept in supporting safety management across a range of situations at the strategic, system and technical levels. The chapter concludes with a discussion on the challenges of the application of the COP concept in major hazards industries.

COP as a mental model in controlling safety

Safety management has traditionally focused on correcting safety concerns, problems or hazards and taking the necessary steps to bring the system back to normal operation. Existing safety approaches seem to rely on what is commonly known as 'feedback control' where feedback guides further corrective actions (Figure 2.1). Therefore, we need to develop a new approach – that not only solves out discrepancies between safety goals and current states – but also helps us understand current operations and risks encountered in industrial facilities. If stakeholders are able to develop a common picture of operations and risks then they are in a better position to anticipate the effects of corrective action and risk mitigation. In order to develop this anticipatory or proactive capability, a safety approach should rely on an 'internal model' of the process that can predict future states of the process and consequences of corrective actions. This type of 'model-driven control' would enable safety analysts to cope with an overload of information and direct attention to critical events in a timely fashion.

In general, the concept of a 'mental model' that drives human performance is well known in ergonomics and human factors engineering (Rouse

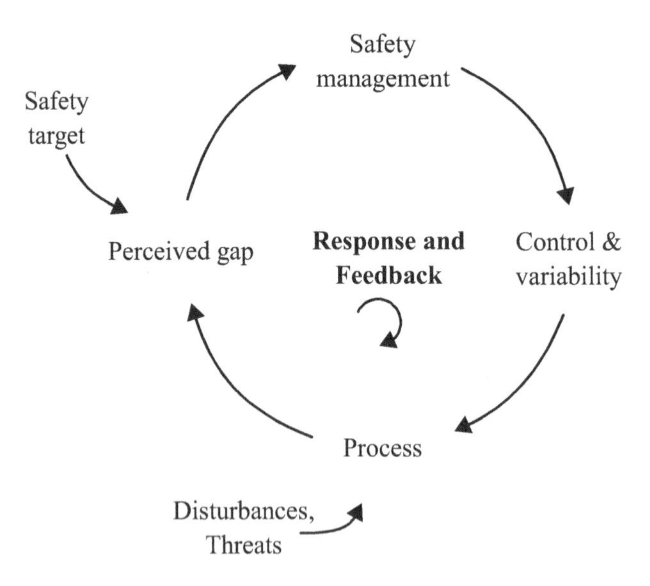

Figure 2.1 A reactive safety management system

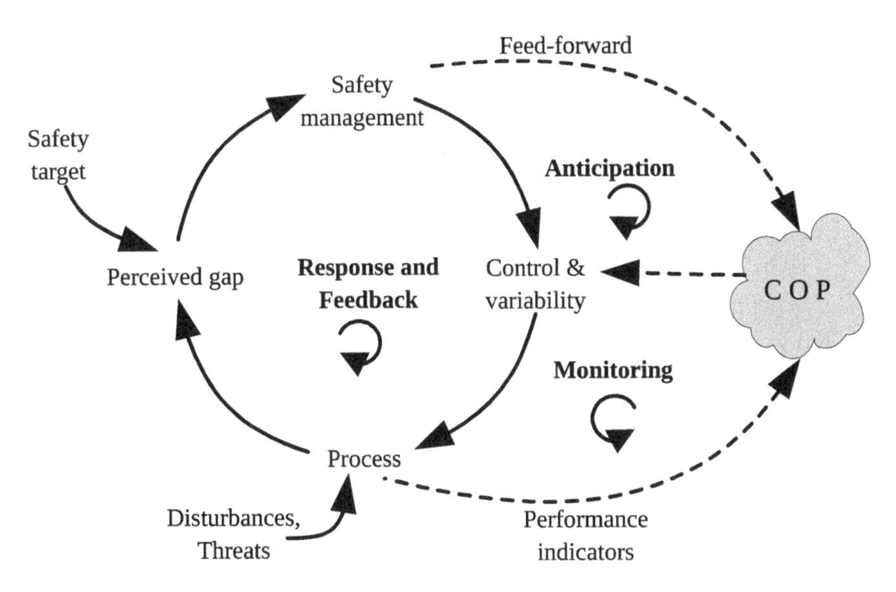

Figure 2.2 A proactive safety management system

et al. 1992; Langan-Fox, Anglin & Wilson 2004)). Despite the long research, however, there has not been a unanimous approach on what should be included within the concept of a mental model. A similar problem may appear when we try to apply the COP concept to examine how managers, supervisors or workers 'understand' the risks in a facility.

Safety practitioners should be able to monitor what could become a threat in the near-term and what could impair their abilities to respond. This monitoring capability can be supported by a common picture of how the technical process works, how people organize their jobs and how the environment affects the process and the people. COP is a mental model that addresses how safety is measured and what Key Performance Indicators (KPIs) can be monitored so that accident precursors are identified and interventions are made before undesired outcomes are ensued. Performance indicators can also provide a good basis for integrating measures of safety with productivity and quality control. Figure 2.2 shows a 'model-driven' safety management system that comprises the four functions of resilience engineering (i.e., respond, monitor, anticipate and learn):

- Given a specified safety target, the safety practitioner has to take control action that changes the technical process in order to produce the desired

output; in turn, this is measured by suitable indexes and feedback mechanisms ('monitor').
- The internal model (COP) must enable safety practitioners to 'anticipate' disturbances so that control actions are taken beforehand. The advantage here is that control actions can prevent adverse events from taking place or intervene before their consequences spread to other parts of the system.
- Safety practitioners should also be able to 'respond' to threats from internal variability due to fatigue of personnel, changes in team composition, unavailability of tools and so forth.
- Safety practitioners must be able to 'learn from experience' which includes modifying their internal model of the process.

There is limited information on what constitutes a COP in a Safety Management System. To some extent, a 'risk registry' can be part of the COP so that all stakeholders and operators are aware of the whole spectrum of risks in an installation. We can view the COP system as a database of risk information that should be accessible to key players but the challenge remains of 'who' should see 'what' information.

Several activities that are critical from a safety or productivity perspective may require a strong coordination between many agents (e.g., safety managers, supervisors, operators and external contractors) as well as communication of information regarding possible side effects, threats and escalation of events. This 'knowledge transformation' process requires that data and information are systematically managed and integrated with people's knowledge of the functioning of the system. Building a common picture of opportunities and threats will allow different agents to understand the systemic causes of safety issues and provide a basis for suggesting practical interventions.

COP systems for emergency response

Emergency response planning is a special part of safety management that deserves further attention because many information technology systems have been developed to support commanders and rescue teams in performing their work in dangerous conditions. Emergency situations are very challenging because operational teams work under high levels of uncertainty and have to make critical decisions using information that may be incomplete, ambiguous or delayed. Emergency incidents are often a race against time, and personnel need to manage themselves so that stress, fatigue and data overload do not undermine their performance. A feature that distinguishes emergencies from other safety critical work is that people are operating under degraded work conditions such as unavailable equipment, shortage of resources, fatigued personnel and hazardous environments. Moreover, emergency events don't 'play by the rules' and people may have to come up with new plans or divert from existing procedures to gain the 'tactical advantage' (Weick & Sutcliffe 2001). For these reasons, many computerized information systems have been developed to help incident command teams acquire a Common Operational Picture and synchronize their efforts despite stress, time pressure and degraded conditions of work (Yang et al. 2009; McMaster & Barber 2009; Chen et al. 2014).

Although many computerized systems are based on the COP concept, they differ with regard to the type of information that is presented to the users and their representation styles. Nevertheless, it is possible to identify some general information requirements for emergency response teams that could provide useful insights on how to transfer the COP concept to a more regular environment in the everyday work of operators in major hazards industries.

According to Seppanen et al. (2013), incident response information covers a wide area of data that can be assigned to the following categories:

- *Baseline information* that depicts the accident type and its location, the time of the accident and the extent of the accident. This is useful to incident commanders in order to define the response teams that should be involved in the emergency response.
- *Static information* concerns the operational environment and includes terrain type and special locations in order to decide on the protected areas and design rescue routes.
- *Information to be created* according to the analysis performed by safety professionals. Examples include: maps of hazards, accessibility to the incident, areas to be evacuated, traffic control and identification of some causes of the incident.
- *Situational information* that is changing and updated constantly. The main information types in this category are hazardous materials, the number

of victims, the resources in use, the location of resources, searching for the missing, the hospitals available, the development of the situation, the responsible crews and changes in weather and visibility conditions.

In many cases, it is important to record the person who provides this information in order to establish a measure of reliability or quality of information. For instance, the general public may provide useful information about the progression of the incident with the use of social media. However, public information may be ambiguous or unreliable and often this requires additional efforts to cross-check and verify this information. Because the portrayed information may be difficult to make sense of, some systems may allow access to related documents and codes of practice. In general, there are numerous ways of combining critical incident information with response checklists, allocation of tasks and emergency plans.

An example of how incident information can become relevant to response teams is shown in Figure 2.3, in the case of an explosion caused by the capsize of a car truck carrying a hazardous material. Figure 2.3 shows a computerized display of critical information in order to support coordination teams in the incident command centre to build a common operational picture.

The central part of the COP system (Figure 2.3) shows a map with the dispersion of fire that is caused by the leak of the hazardous material into two zones of risk. Below the hazmap there are several icons that can be placed on the incident scene to represent the rescue teams, their tools, the fire brigade vehicles, the ambulances and the location of victims. As the incident evolves, the users of the system can move the icons on the hazmap to see how well they control the incident scene. Another part of the screen can be used to portray relevant information provided by the social media as well as watch videos of emergency services in action. On the other side, it is possible to provide decision guidance regarding the structure of the emergency teams, the allocation of tasks as well as the operating procedures that are required to implement an emergency response plan. Of course, it is possible to open and close several windows with other information that may be useful for controlling an emergency.

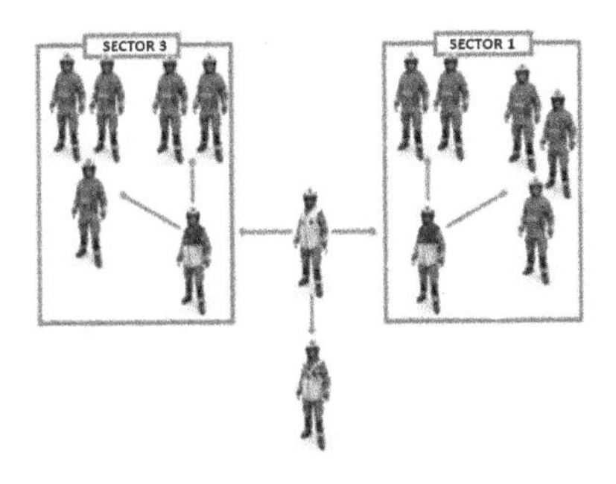

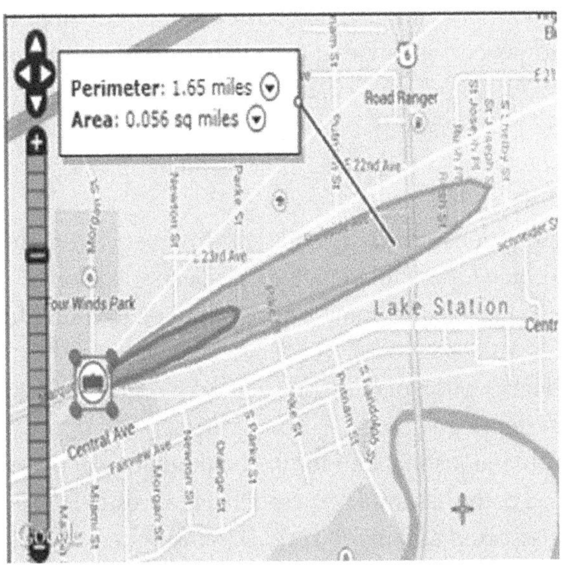

Figure 2.3 A display of critical information in building a COP in emergency response

COP systems for normal and critical safety activities

Operating personnel in major hazards industries are more likely to find themselves working in less severe conditions that may range from routine operations to minor equipment problems and process disturbances. It appears that information requirements for a COP system may change as a function of the situation encountered by the operating teams. Nevertheless, there are some common characteristics that imply that a COP system can be built to cover this range of operations. Most operations are carried out by well trained and motivated professionals, operators have worked together in the past, many hazards can be controlled relatively soon and operating procedures can regulate the coordination of operators and their courses of action.

A useful description of the type and scope of safety information to be provided in a Common Operational Picture has been specified by the NORSOK Standard Z-013. Hence, a common operational picture should provide:

- a balanced description of the scope of risk assessment, the system boundaries and system basis used
- a description and justification for the methodology and tools utilized
- a description of the assumptions used that may result in restrictions to the operations assessed and modifications in the system
- a presentation of the risk acceptance criteria and the results compared with these criteria
- a balanced picture of the contributing factors to risks
- a discussion of uncertainty (i.e., the perspective on risk assessment, the level of uncertainty, and possible unexpected outcomes due to invalid assumptions or insufficient knowledge)
- a discussion of other factors (i.e., divergence of opinion amongst experts or limitations of the modelling) and the robustness of the conclusions with respect to changes in the study basis.

The information specified in the NORSOK Standard Z-013 is not restricted to safety information about hazards, component failures and human errors that is necessary to understand risks at the sharp end of the organizational hierarchy. It also covers information that is useful in performing risk assessment and deciding on a range of safety barriers and controls that would be necessary to recover problems. In this sense, the COP can be useful in supporting the pillars of hazard identification and risk assessment of the Total Safety Management system.

It is also possible to consider how the COP can support the pillar of the safety business case and the pillar of monitoring and learning from the past. It is anticipated that safety information in the form of safety targets, safety performance indicators and monitoring of safety interventions could be included in a COP system. This strategic information could support a better integration of safety interventions and other business processes as well as provide a useful monitoring tool of the effectiveness of safety interventions. In many respects, the COP system can provide a mixture of information and knowledge that is necessary to promote the four pillars of TSM.

Hence, a Knowledge Management System (KMS) can be developed that would provide safety information at three different levels:

- *The strategic level information* regarding safety targets, key performance indicators, incident statistics and work progress indicators
- *The system level information* regarding hazard identification, risk assessment and barrier management
- *The technical level information* regarding hazards, component failures and human errors at the sharp end.

By presenting process safety information into three hierarchical levels it becomes easier to control the problem of information overload as well as decide on the sort of accessibility provided to different users of the COP system (e.g., managers, supervisors and operators). A more elaborate discussion of the three levels of information presentation is made in the following sections.

Strategic level information

Lack of adequate Safety Performance Indicators (SPIs) has been cited as a factor in a number of recent major accidents (Hopkins 2000; HSE 2003, US Chemical Safety and Hazard Investigation Board 2007). In many cases, accidents begin with deviations in the material conditions, in process control, in safeguarding equipment or in human actions from standard procedures that were not detectable because of lack of indications. Against this background and interest from the regulators (HSE 2006), many companies involved in major hazards are developing safety indicator programs. SPIs represent a set of measures focusing on those aspects of organizational performance that are the most critical for the success of an organization.

According to Bellamy and Sol (2012), SPIs developed for process safety performance could serve similar functions to those developed for realizing other business objectives:

- to help a facility monitor and manage its level of safety performance by evaluating its progress towards safety goals
- to give assurance to stakeholders that a plant is being well-managed with respect to major hazards
- to find ways to continuously improve safety.

A set of performance measures will be part of a facility's own monitoring system, providing intelligence about the major hazard control system and how it is performing.

Many Safety Performance Indicators are usually reactive or lagging and measure the results of deviations rather than its precursors (e.g., number of injuries, frequency of leaks, number of equipment damages and so on). A proactive approach however would require industries to monitor information that could potentially lead to incidents and near misses. Examples of leading SPI include indicators regarding safety barriers, administration controls and procedures, operator training, inspections, safety protection devices and so forth. Methods for eliciting SPIs for chemical industries range from checklists of safety management factors (e.g., CCPS 2009) to risk analysis methods such as bowties (e.g., HSE 2006).

In a sense, a dashboard of SPI may be used as an indicator of the effectiveness of a safety management system. Figure 2.4 shows a radar display of a few SPIs that correspond to different pillars of the TSM system. There are 12 spikes that are calibrated so that a normal polygon is formatted when the SPIs take their maximum values. The analysts have to decide the levels that indicate acceptable, tolerable or unacceptable risks so that

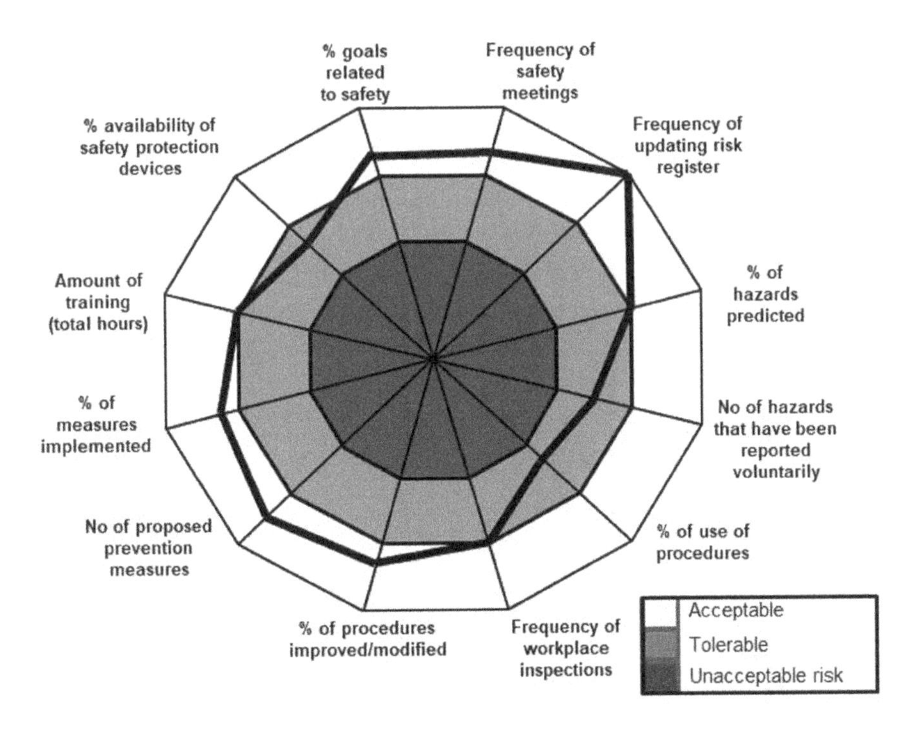

Figure 2.4 A radar display where SPI (spikes) are calibrated to produce a normal polygon

the polygon is separated into three risk levels. A distortion of certain critical SPI that fall within the tolerable area may act as early warnings that should be attended to in order to prevent escalation of the risk into the unacceptable area.

As the list of SPI increases, it becomes more difficult to monitor their individual trends. Instead, it could be possible to organize the SPI in several functional groups and create collective indices or overall measures of safety performance. A methodology for aggregating multiple SPI into suitable measures of performance is presented in Chapters 7 and 8.

Systems level information

The requirements for a COP specified in the NORSOK Standard Z-013 mainly refer to the pillar of risk management of the TSM approach. In this sense, the COP system should provide the necessary data for performing risk analysis regardless of the chosen method (e.g., bowties, event and fault trees, cause-consequence models). To achieve this, the COP system should help analysts to identify the threats, hazards, accident sequences, barriers and proximal or distal factors that affect the likelihood of accidents. In particular, the COP system should equip analysts with knowledge about technical systems, human-system interactions and operations, reliability issues and factors affecting performance.

The COP concept can be the basis for developing a Knowledge Management System (KMS) for exploring system knowledge and managing safety barriers. The knowledge base may contain information about many aspects such as:

- Plant equipment and systems
- Failure modes and hazards
- Standard operating procedures and work permits
- Competency and accreditation requirements
- Linkage to safety rules and regulations.

The knowledge database could be linked to a risk model of bowties that specifies prevention barriers and mitigation barriers. Chapter 9 proposes a Computerized Barrier Management System (CBMS) for a chemical manufacturing plant. Figure 2.5 shows a display of the CBMS with a description of tasks, roles, procedures, critical activities and other related operations. At the top of the screen there is a log of safety critical activities related to the unloading of hazardous materials from a car truck into a storage vessel in a major hazards facility. The user can select each particular phase of unloading (e.g., prevent sparks and fire propagation) and examine the work factors and barriers in place that influence the reliability of a phase

The system registers safety data about: the work organization (e.g., the personnel who carry out the critical activity), the accreditation models, the barriers for preventing sparks and fires (e.g., gas detectors, alarms, grounding in tanks, other maintenance and inspection tests and procedures for firefighting). There is also additional information about appropriate procedures for carrying out the unloading tasks, work permits, restrictions and equipment involved in this critical activity. By clicking on each barrier, at the right bottom part of the screen, the user can get access to proper documents that have been issued by the safety department (i.e., technical level information).

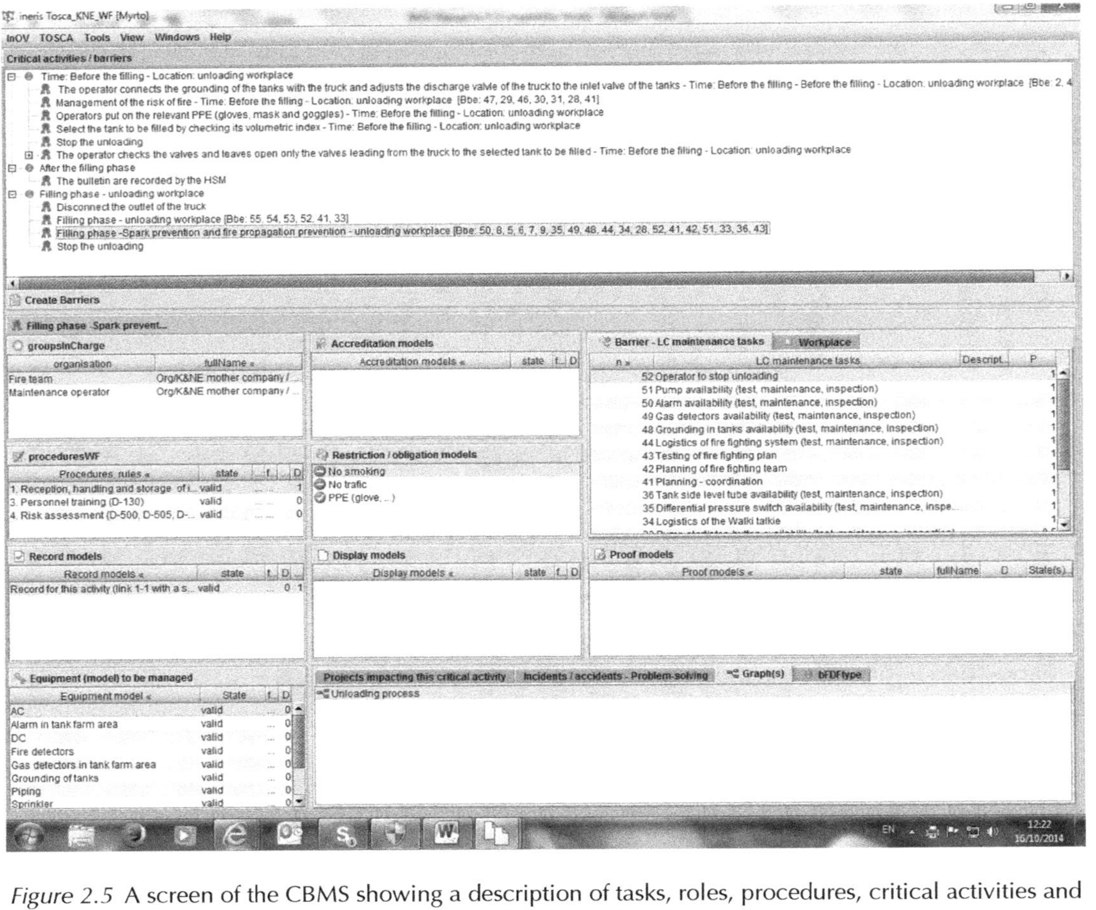

Figure 2.5 A screen of the CBMS showing a description of tasks, roles, procedures, critical activities and other related operations

Apart from having in one place all the necessary knowledge about risk analysis, the Knowledge Management System in Figure 2.5 allows operators and system users to update safety information resulting from changes in equipment, procedures and control systems (for more information see Chapter 9). Hence, existing risk analyses can be retrieved and updated to reflect the current state of affairs with a small amount of effort. The system can make the risk assessment a 'live' process rather than a 'static' one that is performed only during the initial stage of plant commission.

Technical level information

Most users of the COP system interact at the technical level, by providing information about work documents, safety data sheets, safety restrictions, equipment failure data, reported human errors, warnings and points of concern and data about safety devices. A wide range of users can be included at this level in order to provide this technical information or update existing information, issue warnings and concerns, publish the results of near misses and incidents and so on.

Figure 2.6 shows a display of the Computerized Barrier Management System (CBMS, in Chapter 9) that provides the following safety information: (*a*) Material Safety Data Sheets, (*b*) documents of any type such as texts, photos, PID diagrams for the storage tank, (*c*) procedures for reception, handling and storage of incoming material, equipment maintenance, personnel training, risk assessment and unloading, (*d*) restrictions and obligations in certain areas, such as no smoking, no traffic and use of personnel protective equipment, (*e*) safety equipment such as the firefighting system in the storage area (foam tank, sprinklers and gas detectors) and all other equipment required for the operation of the storage tanks.

The user can interact with the technical information in Figure 2.6 by retrieving the relevant work document to seek advice or modify them when there are changes or modifications. In addition, it may be possible to retrieve Pipe & Instrumentation Diagrams and equipment 2D or 3D pictures that are related to the application of a work document. In this sense, 2D or 3D equipment pictures can be placed side by side to work instructions to facilitate the evaluation of them. Other means of portraying technical information are presented in Chapter 9.

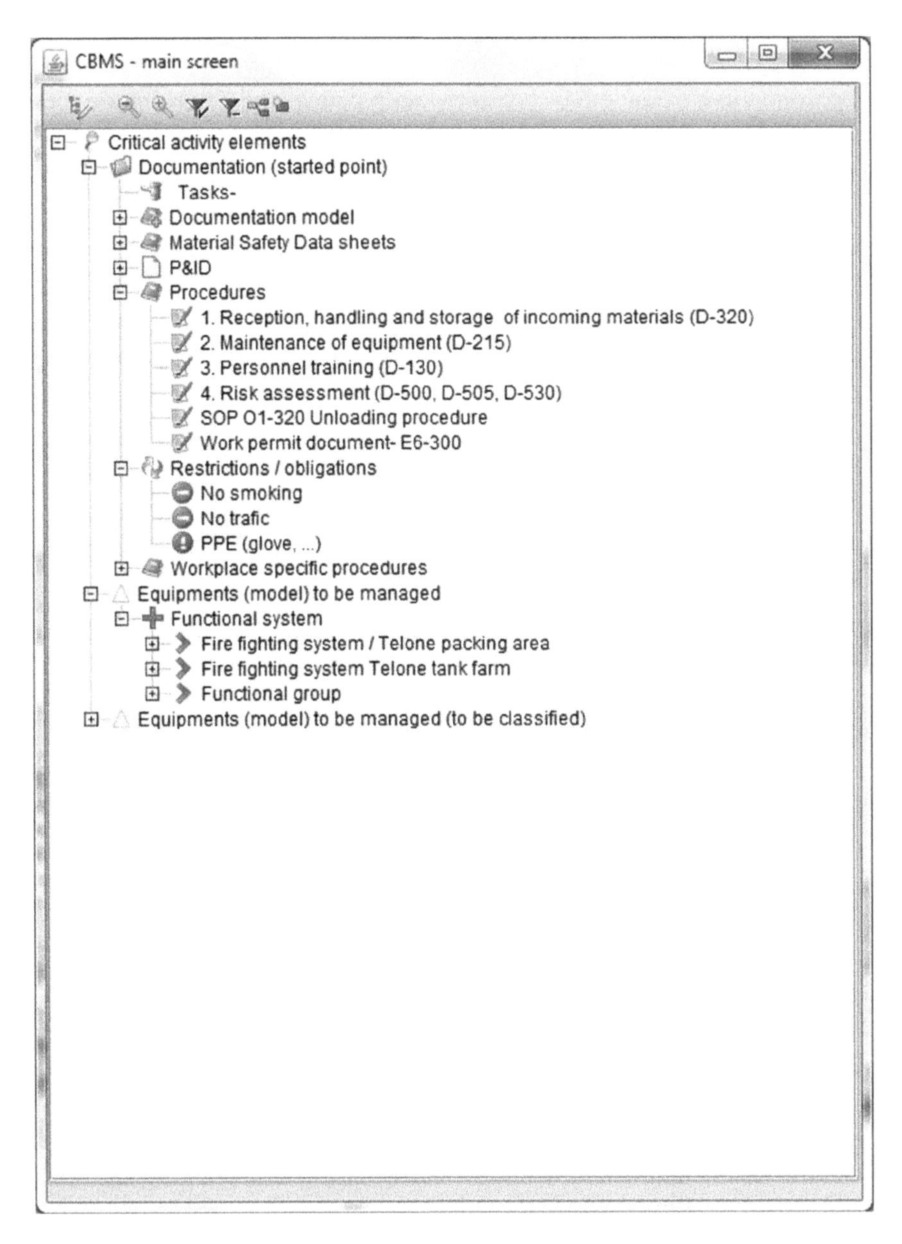

Figure 2.6 A menu with a list of documents regarding equipment, people, safeguards and procedures at the technical level

Concluding remarks and challenges in building a COP system

The Common Operational Picture provides practical and timely information about work hazards and threats that may escalate and become uncontrollable. The amount of safety information to be collected and displayed on a central system remains a significant challenge. The risk is that a Knowledge Management System that structures huge amounts of safety data can become overwhelming for safety practitioners and managers. The strategic, system and technical levels of interaction offer some structure in a complex KMS and allow different users to interact with the system at their preferred levels. For instance, managers can interact at the strategic level, supervisors at the system level whilst operators at the technical level. The weakness of this approach may be that operators may feel that they only contribute information to the system without having access to some feedback at the strategic or system levels. Hence, there is a need for a mixed mode of operation that allows operators and supervisors to use the system at the two higher levels of interaction. The particular scope of the mixed mode of interaction should be decided on the basis of several criteria related to the size of the company, its processing materials, the safety policy and the particular culture of the company.

Another challenge for a KMS that implements the COP concept regards the policies and procedures for its initial design, especially: who will be involved in the design process, what procedures should be used to collect safety information, how to interface a KMS with other existing information technology systems and so on. It is anticipated that different people shall be involved at the three levels of interaction but the amount of data required will depend on the scope of the system. Difficulties at this step may include operators and supervisors having different perceptions and methods for their critical activities; this lack of standardization could create problems at determining what is best practice for performing several operational or maintenance tasks. The design procedure is also a challenge since most of the safety data required may be in a paper format or may not be available at all; hence, the design of the system may impose extra work for generating safety data or transforming them into an electronic format. Lastly, a KMS should be integrated with existing information technology systems and there may be overlaps of safety information among the various systems; hence, deciding how to apportion safety information among different systems can be an important challenge.

The following chapters in the book provide a variety of design applications of the COP concept from risk registers to Computerized Barrier Management Systems. Another two chapters have drawn on the role of SPIs in safety management and provide methods for aggregation into overall safety incidents.

References

Bellamy, LJ & Sol, VM 2012, 'A literature review on safety performance indicators supporting the control of major hazards', *RIVM Report 620089001/2012 National Institute for Public Health and the Environment*. The Netherlands.

CCPS (Center for Chemical Process Safety) 2009, *Guidelines for process safety metrics*, Wiley. Hoboken, New Jersey.

Chen, T, Su, G & Yuan, H 2014, 'Creating common operational pictures for disaster response with collaborative work', *WIT Transactions on Information and Communication Technologies*, vol. 47, pp. 393–400.

Hopkins, A 2000, *Lessons from longford*, The Esso Gas Plant Explosion, CCH Australia Ltd., Sydney.

HSE (Health and Safety Executive) 2003, *BP grangemouth: Major incident investigation report*, www.hse.gov.uk

HSE (Health and Safety Executive) 2006, *Developing process safety indicators: A step-by-step guide for chemical and major hazard industries*, HSE, Sheffield.

Lagan-Fox, J, Anglin, J & Wilson, JR 2004, 'Mental models, team mental models, and performance: Process, development, and future directions', *Human Factors and Ergonomics in Manufacturing*, vol. 14, no. 4, pp. 331–52.

Liu, B, Wu, CH & Liu, HJ 2011, 'Research on the human factors of common operational picture', *Procedia Engineering*, vol. 24, pp. 182–286.

Luokkala, P, Nikander, J, Korpi, J, Virrantaus, K & Torkki, P 2017, 'Developing a concept of a context-aware common operational picture', *Safety Science*, vol. 93, pp. 277–95.

McMaster, R & Barber, C 2009, 'Multi-agency operations: Cooperation during flooding', in D De Waard, J Godthelp, FL Kooi & KA Brookhuis (eds.), *Human factors, security and safety* (pp. 1–15), Shaker Publishing, Maastricht.

NORSOK Standard Z-013 2001, *Risk and emergency preparedness analysis*, www.standard.no/pagefiles/955/z-013.pdf

Rouse, W, Cannon-Bowers, J & Salas, E 1992, 'Role of mental models in team performance in complex systems', *IEEE Transactions on Systems Man and Cybernetics*, vol. 22, no. 6, pp. 1296–308.

Seppanen, H, Makela, J, Luokkala, P & Virrantaus, K 2013, 'Developing shared situational awareness for emergency management', *Safety Science*, vol. 55, pp. 1–9.

Wahlstrom, B & Rollenhagen, C 2012, 'Safety management: A multi-level control problem', *Safety Science*, vol. 69, pp. 3–17.

Weick, KE & Sutcliffe, KM 2001, *Managing the unexpected*, Jossey Bass, San Francisco.

Yang, L, Prasanna, R & King, M 2009, 'On site information systems design for emergency first responders', *Journal of Information Technology and Application*, vol. 10, no. 1, pp. 5–27.

US Chemical Safety and Hazard Investigation Board 2007, *Investigation report refinery explosion and fire*, BP Texas City, www.chemsafety.gov/

PART

2

Understanding hazards and risks

Process risk assessment

From the basics to new frontiers

Micaela Demichela and Gabriele Baldissone

Introduction

Process risk assessment is nowadays a mature discipline that is based on consolidated techniques and tools for each phase of its cycle. The definition of "risk" and the methodologies falling under the acronym PRA – Probabilistic Risk Assessment – have arisen from the nuclear domain and laid the foundations of technological risk evaluation (Rasmussen 1975).

Technological risk refers to *the undesired consequence of a particular hazardous activity in relation to its likelihood of occurrence*. This clearly means that the risk (R) can be seen as the product of two parameters: F, the expected probability that a hazardous event could occur within a given time interval; and M, the severity of the consequences ensued (Magnitude).

From the Rasmussen's Report the logical link between the two is set to:

$$R = F \cdot M$$

This simple definition of technological risk allows safety analysts to assess risks and make comparisons between the foreseeable risks associated to a given process or plant or to examine whether risks fall within particular tolerability limits.

Since process plants are usually complex socio-technical systems, the risk assessment process requires a procedure to control this complexity and to support risk-based decision making within the Total Safety Management framework (see Chapter 1).

The risk assessment procedure

As discussed in Chapter 1, standard ISO 31000:2009 defined a framework for risk assessment that is coherent with most process risk assessments from early days, although with some variations. According to Figure 3.1, the phases of process risk assessment can be summarized as follows:

1 Identification of hazards
2 Risk assessment, taking into account the probability of occurrence and the consequence severity
3 Risk management and control.

It is worth noticing that in process risk assessment the initial concern is about hazards, intended as the inherent characteristics of a process, material or equipment that could cause damage to operators, assets or environment. The first step in process risk assessment in thus a qualitative one.

Available methodologies that support the risk assessment phases are classified according to their outcomes:

- Qualitative (Q) assessments allow the identification of possible unwanted events or potential accidents as well as their causes and consequences, without information on the entity of the risk

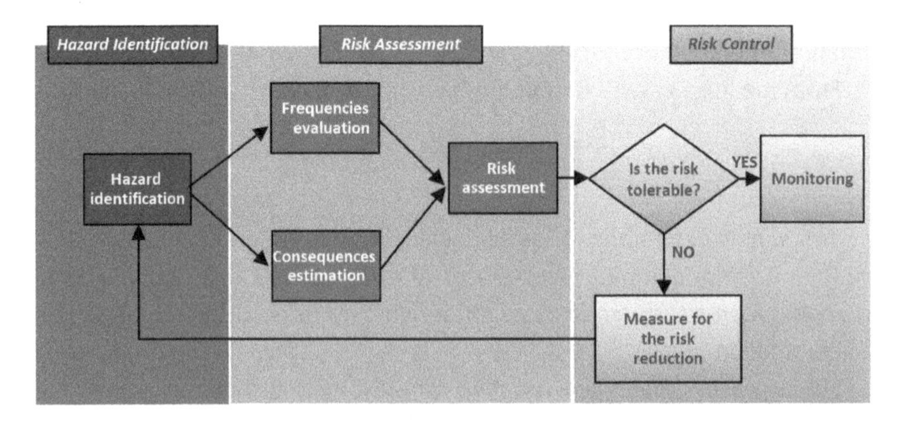

Figure 3.1 Risk assessment and management procedure

- Semi-quantitative (SQ) assessments produce a risk index that allows the estimation of the risk, although in an indicative way
- Quantitative (QRA) assessments calculate the probability of occurrence of a hazardous event and the consequent damage, giving a numerical estimation of the risk.

The choice of the level of detail of the analysis depends on the level of risk associated to a plant (Figure 3.2) and should take into account regulatory constraints (e.g. for process plants under "Seveso" regulation, the fully quantitative risk assessment is usually required). When higher risk complex systems are concerned, it is allowed to carry out a semi-quantitative (SQ) pre-screening able to highlight the riskiest situations to be further analyzed with fully quantified – and more resource consuming – risk assessment methodologies (QRA).

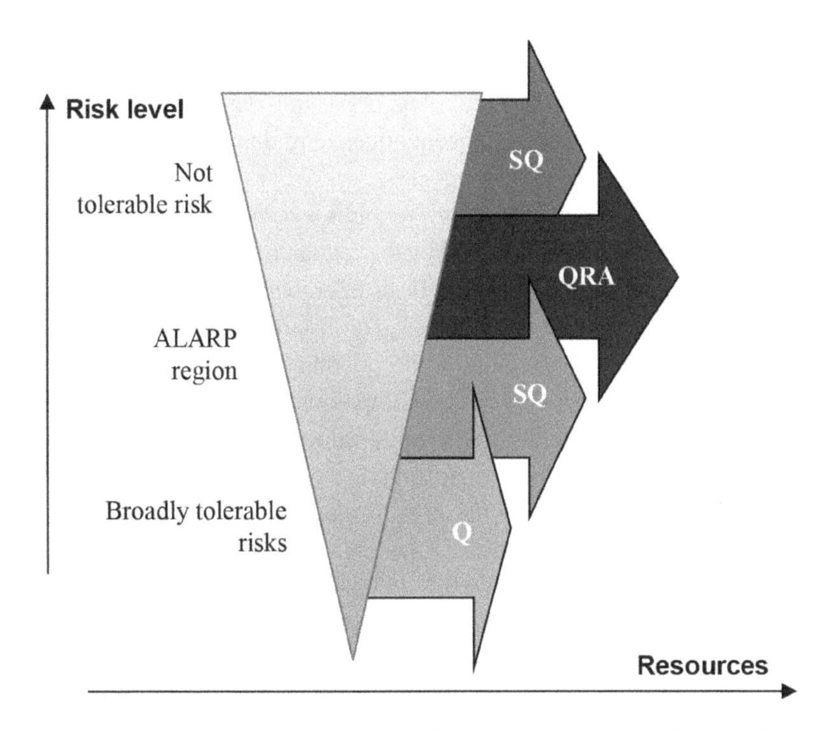

Figure 3.2 Proportionality between the risk level and the level of detail of the risk assessment methodologies

The most commonly used methods for hazard identification in process industry include:

- Checklists and What If? Analysis
- Failure Modes, Effects and Criticality Analysis (FMECA)
- Hazard and Operability Studies (HAZOP).

Traditional methods for probability quantification are:

- Event Trees (ET)
- Fault Trees (FT).

More recently, the bowtie methodology started to be a state-of-the-art method that also allows a certain degree of risk quantification (see Chapter 4).

The estimation of the severity of consequences estimation is a complex activity that varies according to the type of accident or unwanted outcome considered (e.g. release of energy or matter, loss of production), the type of desired outcome (e.g. damage mapping, economic cost of the malfunction), and the type of hypothesized consequences (e.g. damage to persons, damage to structures, economic damage).

In the end, the phase of risk control requires a comparison of the risk value with the tolerability criteria provided by the regulations or chosen by the company (if more restrictive than legal limits). Tolerability criteria depend on the level of detail of the risk assessment itself. For instance, semi-quantitative evaluations will rely on risk matrixes, while quantitative risk assessment returns precise numerical outputs that can be compared to existing risk thresholds. In risk matrixes, both probability and severity values are discretized in intervals; their intersection defines a risk map (matrix), where at least three areas are identified: (i) the non-tolerable risk area (in dark grey), representing the risks that must be reduced through preventive and/or protective measures; (ii) the ALARP area – As Low As Reasonably Practicable – (in light grey), that represents the risks to be reduced according to a cost-benefit prioritization; and (iii) the tolerable risk area (in grey), representing the risks that do not require further actions for reduction. A similar subdivision of the risk space is done in case of numerical risk thresholds. Figure 3.3 shows an example of the two criteria for tolerability assessment: Figure 3.3a represents a risk matrix applicable in industrial domains with various impacts on operators and environment; Figure 3.3b represents the individual risk tolerability adopted in The Netherlands.

(A)

DAMAGE CODE		CONSEQUENCES		PROBABILITY				
		Personnel	Environment	1 Any occurrence in industry is unknown or appears unlikely	2 Has occurred in the industry	3 Has occurred within the company sector	4 Has occurred within the operating company	5 Can occur in the company several times a years
Minimal	1	Negligible injuries/ illness	Negligible impact					
Low	2	Minor injuries/ illness	Minor impact					
Medium	3	Major injuries/ illness	Locally limited impact					
High	4	1 to 3 fatalities	Major impact					
Very high	5	Several fatalities	Massive impact					

Figure 3.3 Examples of the tolerability criteria adopted in process risk assessment

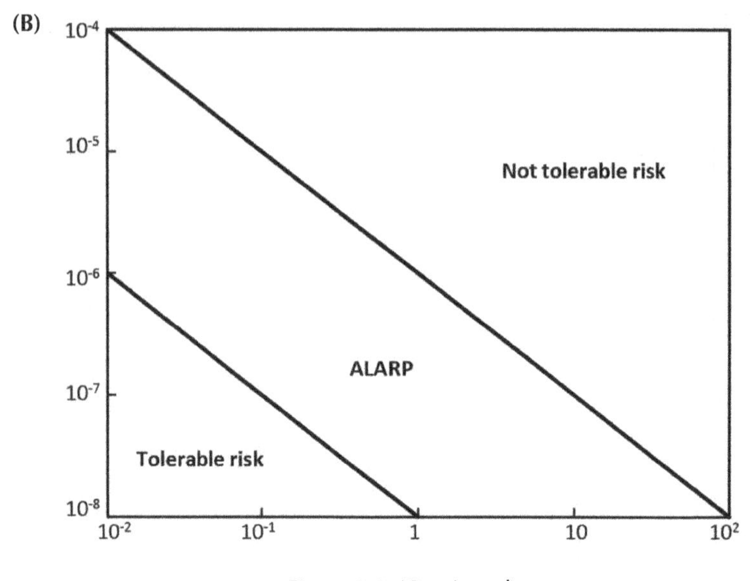

Figure 3.3 (Continued)

Hazard identification

Some common methods for "hazard identification" are described in the following paragraphs; they identify possible hazards and their possible consequences. The choice of the method depends on the degree of available knowledge about the process and the type of desired results.

"What if?" analysis

"What if?" analysis is a brainstorming-based approach that investigates on the consequences that could ensue in case a certain hazardous event occurs. Since the methodology is not well structured, it can overestimate or underestimate some aspects of the events or lead to an incomplete identification of hazards.

Better results can be achieved when this methodology is coupled with a checklist: in this case, "what if?" analysis checks the behaviour of the system and, at the same time, the checklist gives a reasonable assurance of completeness.

Checklist

The checklist method is very commonly used as it is made of a set of questions with a "yes" or "no" answer. The number and order of the affirmative or negative answers allows analysts to identify hazards. The results obtained by this method depend mainly on the degree of knowledge that analysts have about the process.

The quality of the checklist depends on the experience of the person who prepares and applies it; however, literature provides several examples that could be adapted to the specific use.

The sole use of the checklist is suitable only for simple cases or plants with well-known technology. In other cases, the use of the checklist can be useful to collect information as a basis for subsequent complex techniques.

Failure Modes Effects and Criticality Analysis

The Failure Modes Effects Analysis (FMEA), or its variation, Failure Modes Effects and Criticality Analysis (FMECA), have become two of the most used techniques for the identification of hazards in complex systems.

This methodology comprises the identification of all the potential failure modes of the equipment, with a consequent investigation of their possible causes and effects on the system. The analysis procedure is encoded in a devoted technical standard, the IEC 60812:2006 Analysis techniques for system reliability – procedure for failure mode and effects analysis (FMEA).

FMECA is an enhancement of FMEA that incorporates a criticality analysis. Criticality is a function of the severity of the effect and the frequency with which it is expected to occur. The criticality analysis involves assigning to each failure mode a frequency and to each failure effect a severity.

The purpose of FMEA is to identify the failures which have undesired effects on system operation. Its objectives include: (i) the identification of failure modes, of the sequence of events associated with them and of their causes and effects; (ii) the classification of failure modes through relevant characteristics, including detectability, diagnosed capability, testability, item replaceability, compensating and operating provisions; and, for FMECA, (iii) an assessment of the criticality of each failure mode.

The main documentation used in FMEA is the functional diagram, or a reliability block diagram.

The failure modes may be described at two levels: (i) generic failure modes such as failure during operation, failure to operate at a prescribed time, failure to cease operation at a certain time and premature operation – and (ii) specific failure modes – as cracked/fractured, distorted, undersized, etc.

The failure causes associated with each mode can be classified into various groups: (1) specification, (2) design, (3) manufacture, (4) installation, (5) operation, (6) maintenance, (7) environment and (8) uncontrollable forces.

The failure effects may involve changes in the operation, function or status of the system and these should be identified by the analysts. Failure effects can also have local or end effects. Local effects refer to the consequences at the level of the element under consideration and end effects to those at the highest level of the system.

FMEA is recognized to be an efficient method of analysing elements which can cause failure of the whole, or of a large part, of a system. It works best where the failure logic is essentially a series one. It is much less suitable where complex logic is required to describe a system failure.

FMEA is an inductive method. A complementary deductive method is provided by Fault Tree analysis that is more suitable for the analysis of complex failure logic.

Hazard and operability studies

The more commonly adopted method of hazard identification in the process industry is the Hazard and Operability Analysis (HAZOP). This type of analysis is carried out by a multi-disciplinary group with the purpose of reviewing the entire production process, in order to identify possible hazards and potential problems using an organized approach based on guide words. This method is based on the analysis of the potential deviations of process variables from the intended design and their operational values. HAZOP was developed in the sixties of the 20th century by ICI for the analysis of critical situations and underwent several adaptations since then.

This type of analysis can use a flow diagram (PFD) or a detailed piping and instrument diagrams (P&ID) to examine deviations from normal conditions. The completeness of the analysis strongly depends on the choice of the level of project development to which the technique is applied: when HAZOP is performed at an early stage, some information may still be missing and, at a more advanced stage, any corrective action would be very expensive.

The basic concept of HAZOP analysis is the investigation, for each part of the process, on the possible types of deviations from design conditions, and on the possible causes and consequences.

Deviations can regard all aspects and parameters involved in the process such as materials, activity, equipment, source, destination, time and space.

Guide words have to be applied to the process parameters in any node of the plant they are relevant, for guaranteeing a systematic analysis. This coded system helps analysts to find and analyse the deviations that may be present in the system. A basic set of guide words includes:

- NO or NOT, Negation of intention
- MORE, Quantitative increase
- LESS, Quantitative decrease
- AS WELL AS, Qualitative increase
- PART OF, Qualitative decrease
- REVERSE, Logical opposite of intention
- OTHER THAN, Complete substitution.

Depending on the different variables, variations of the guide words may be applied to make sense; e.g. the MORE or LESS can be applied in case of durations or frequencies, but in case of absolute times SOONER and LATER may be more suitable.

In order to grant the analysis to be systematic and complete, a procedure has been developed, as summarized in Figure 3.4 (Mannan 2005).

Similarly to other hazard identification techniques, HAZOP is an activity based on teamworking. The choice of team members is one of the most delicate matters of the analysis, because it influences the points on which the attention will be focused. Therefore, the choice of the group depends on the purposes of the analysis: review of the design, identification of major hazards, process or operation optimization and so on.

A typical team includes the study leader (facilitator), a project engineer, a process engineer, an instrument engineer and the commissioning manager. Other personnel who are often included, depending on the nature of the project, are a chemist, a civil engineer, an electrical engineer, a materials technologist, an operations supervisor, an equipment supplier's representative, etc.

The team members provide the technical input in response to the guide words and amplify the information about the plant design given in the plant diagrams, operating instructions, etc.

The final result of the analysis is a series of tables presenting the outcomes obtained, where the possible deviations with their possible causes and consequences are evidenced.

Two main limitations correlate to this kind of analysis: the first one arises from the assumptions of the method that the design of the system is

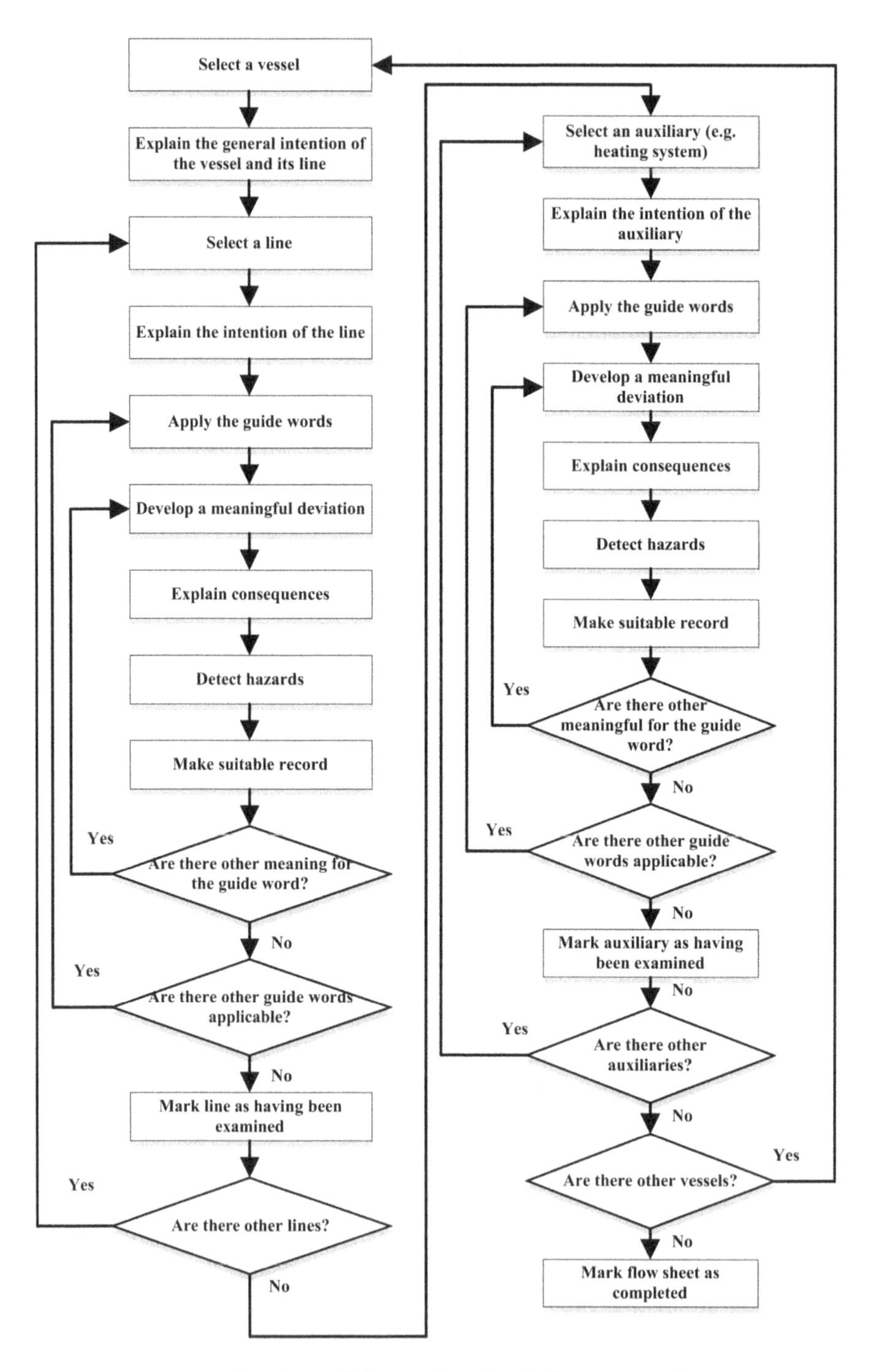

Figure 3.4 Hazard and operability studies: detailed sequence of examination

in accordance with the appropriate codes and requirements. The second limitation is inherent in the method: HAZOP is static and does not take into account transient events. Therefore, it is not particularly well suited to deal with spatial features associated with plant layout and their resultant effects.

In more recent years, different types of HAZOP have been developed to respond to different situations and to simplify certain steps of the analysis. One of the more interesting variations has been proposed by Piccinini and Ciarambino (1997) in an effort to facilitate the subsequent development of the logic trees for probability estimation. The proposed Recursive Operability Analysis (ROA) differs from the traditional analysis in the following terms:

1 The way the relevant deviations are defined, not through the use of guide words, but through expert judgement;
2 Two recursive mechanisms are proposed, in order to develop the consequences until the TOP EVENTs and to deepen the causes until the primary causes, not just in one row, but following their evolution in time and along the process;
3 The way of recording the outcomes of the analysis that allows analysts to build almost automatically a Fault Tree.

(Piccinini & Ciarambino 1997)

This type of analysis proceeds according to the flow chart in Figure 3.5: the first activity consists of the subdivision of the system into smaller subsystems. The next stage concerns the identification of the nodes, which are points where deviations of the process variables (temperature, pressure, etc.) may develop or propagate. For each node, the process variables that may be subject to deviations have to be identified.

Then the analysts can proceed with the analysis of causes and consequences for all deviations, according to the mechanisms in Figure 3.5. and recording the results in an encoded table. Figure 3.6 shows an application of the analysis to a simple case study of a frying system, that evidences the links between ROA tables and Fault Trees (Colombo & Demichela 2008).

It appears that ROA partially overcomes HAZOP limitations in terms of the analysis of time and spatial propagation of the process deviations.

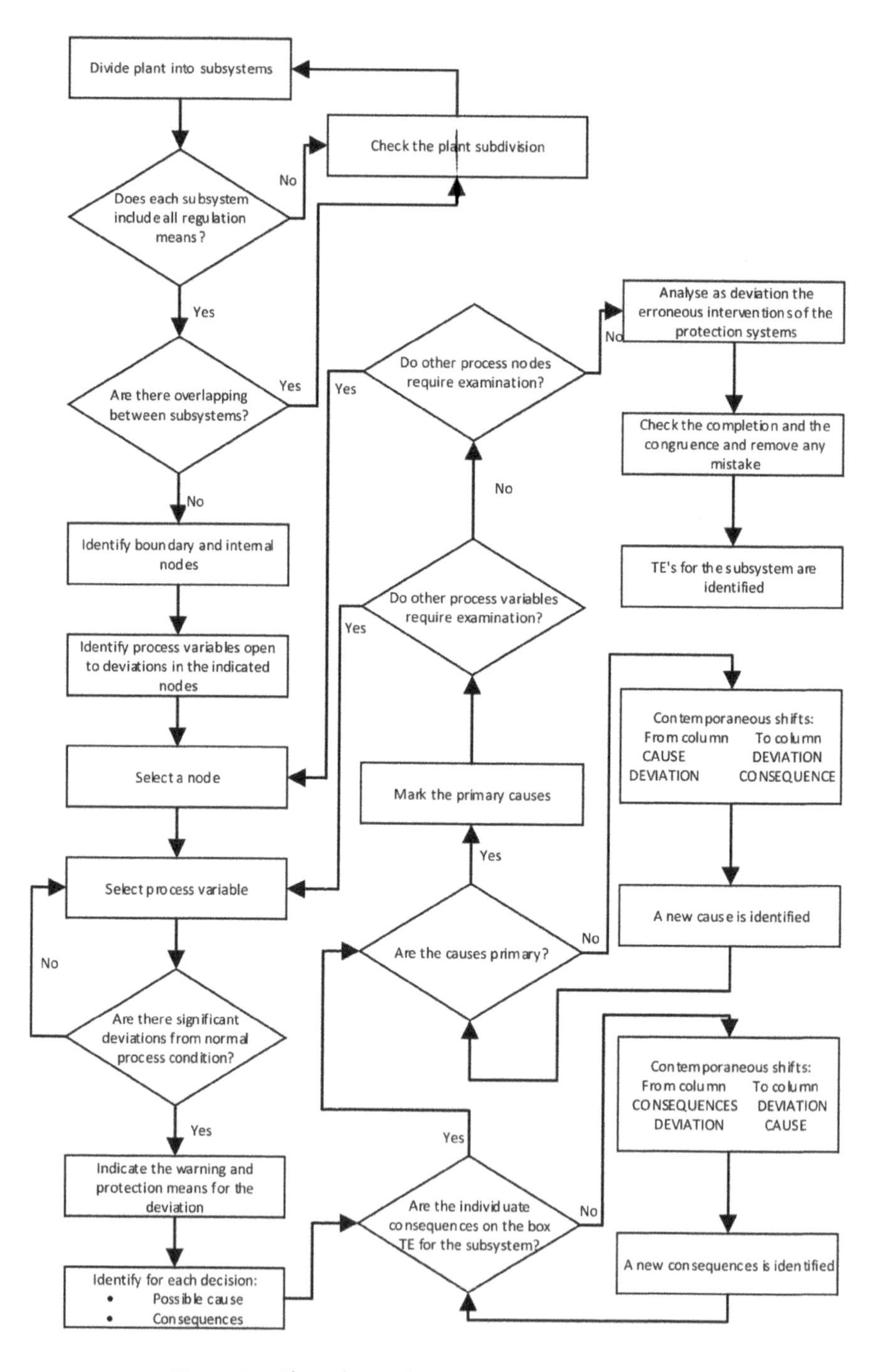

Figure 3.5 Flow sheet of recursive operability analysis

Source: Piccinini & Ciarambino 1997

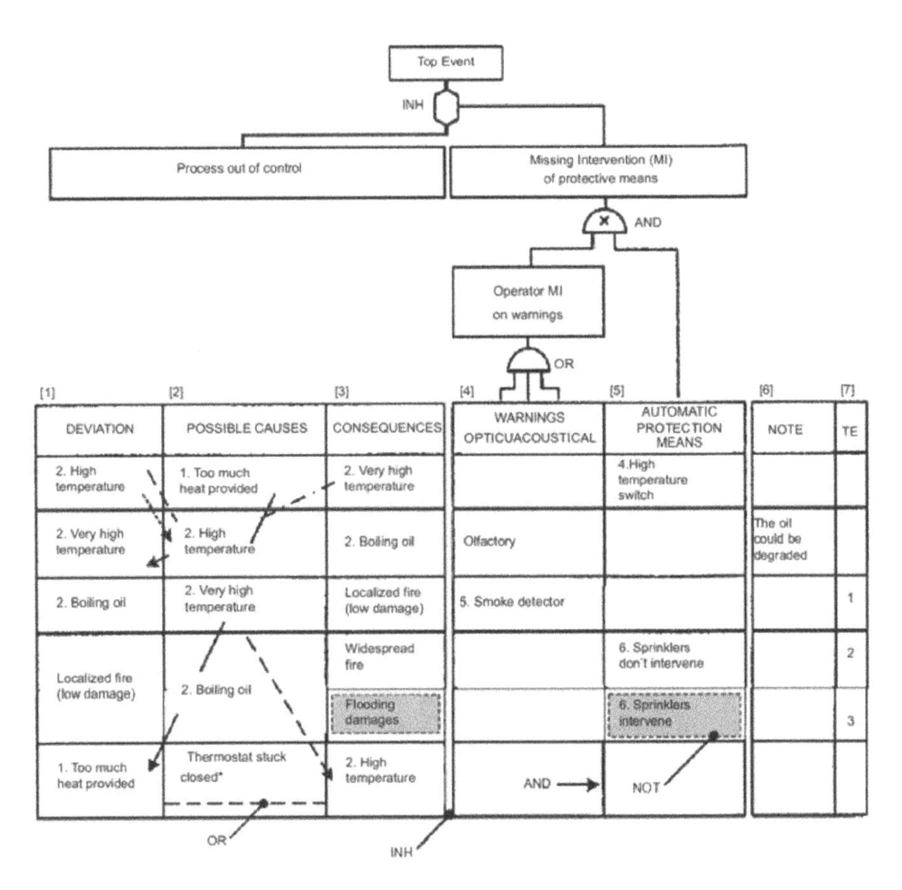

Figure 3.6 ROA tables and inherent links to Fault Trees

Source: Colombo & Demichela 2008

Estimating the probability of occurrence

This section describes two techniques commonly used to estimate the probability of occurrence of unwanted events, identified in the previous stage: Event Trees (ET) and Fault Trees (FT).

The ET is a graphical structure that represents the possible consequences that may arise from a single initial event.

The FT is a graphical representation of all the possible causes for a given event and the logical links between these causes.

More recently, the bowtie analysis has been introduced; it links FT and ET to better address the identification and availability requirements of protection systems (i.e., barriers to deviations or consequence propagation).

Event trees

This technique can be both qualitative and quantitative. In its simplest form, an Event Tree (ET) is a graphical representation that puts in evidence the possible consequences from an initial event. The quantitative form allows the analysts to evaluate the probability (or frequency) of potential consequences.

An example of the graphical structure is shown in Figure 3.7, with the starting event on the left, from where the different branches are built. The upper branch represents the occurrence of the event described at the top of the column while the lower branch represents the non-occurrence.

In process risk assessments, the ETs are used to identify and/or quantify the probability of occurrences of the possible consequences derived from a Top Event.

Figure 3.7 represents the branches of an ET used for quantitative purposes for an initiating event with a certain probability of occurrence (events/year). The consequences arise due to the contemporary occurrence of the events that constitute the branch from the initiating event to the particular consequence. Hence, the final probability of a consequence can be calculated multiplying the probabilities of the initiating event with those of intermediate ones in the particular branch.

Fault tree

The Fault Trees (FT) are logic trees representing the logical links between a Top Event and all its possible causes, in a top-down diagram.

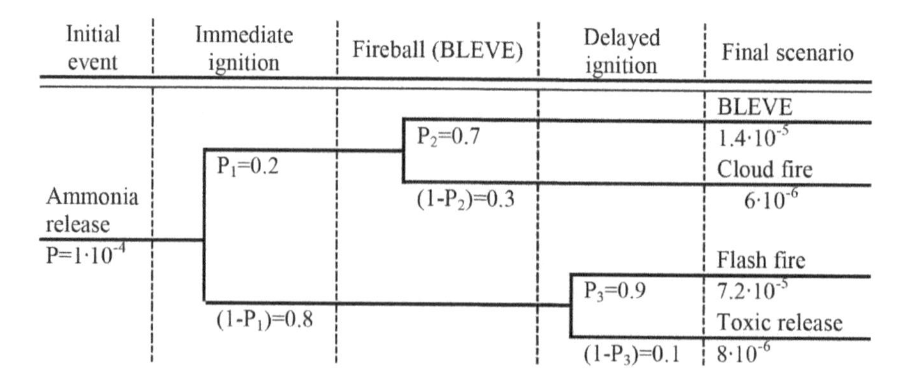

Figure 3.7 Example of a quantified Event Tree

The representation adopts a conventional symbology both for the events (primary, intermediate and top) and for the logical relationships among them (logic gates).

A summary of symbols, as listed in literature, is shown in Figures 3.8 and 3.9.

The Fault tree methodology can be used to describe all the sequences that may lead to a certain event (qualitative), and to estimate the probability of occurrence of the Top Event, starting from the probabilities of the primary events (quantitative).

In this type of analysis, a device can only assume two states: the state of correct functioning or failure. This is a simplifying assumption since the operation of equipment can have many intermediate states that are neglected by the analysis. On the other hand, the use of this approximation allows analysts to hypothesize a binary behaviour and apply Boolean logic, which simplifies the numerical calculation of the Fault tree.

An essential element for the construction of the Fault Tree is the definition and the understanding of the system. Both the system itself and its boundaries need to be clearly defined. Information about the system is generally available in the form of functional diagrams such as

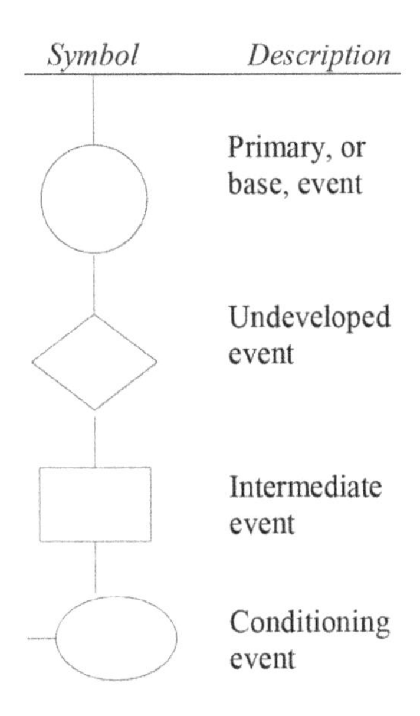

Figure 3.8 Fault tree event symbols

Symbol	*Description*
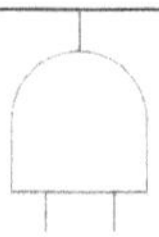	AND gate: output occurs only if all the inputs occur
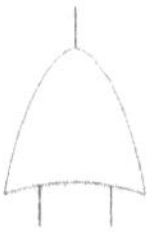	OR gate: output occurs if one or more inputs occurs
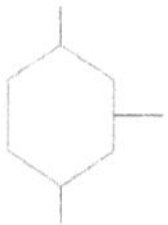	INHIBIT gate: output occurs if the input occurs in presence of the specific enabling condition (the event to right of the gate)
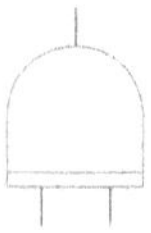	PRIORITY AND gate: output occurs if all inputs occur in a specific sequence
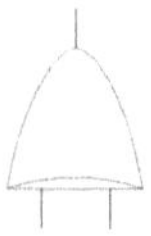	EXCLUSIVE OR gate: output occurs if one, and only one, input occurs
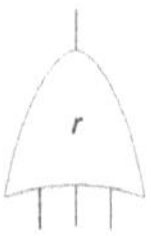 r	VOTING gate: output occurs if there exist r-out-of-n inputs
	TRANSFER IN: symbol indication that the tree is developed further at the corresponding TRANSFER OUT symbol
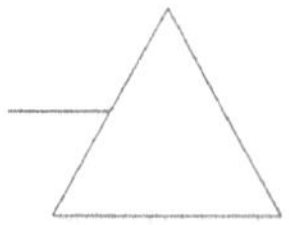	TRANSFER OUT: symbol indication that the portion of the tree below the symbol is to be attached to the main tree at the corresponding TRANSFER IN symbol

Figure 3.9 Fault tree symbols for logic gates

Source: Mannan 2005

piping and instrument diagrams and more detailed instrumentation and electrical diagrams. In complex, not standardized, process plants, the Fault Trees should be developed starting from an HAZOP analysis, which allows analysts to identify the critical events and their links with particular causes.

For a process system, the top event may be a failure mode of a piece of equipment while the immediate causes may involve the failure mechanisms for that particular failure. In turn, these constitute the failure modes of the contributing subsystems, and so on.

The development of a Fault tree, when not guided by a HAZOP, is a creative process, as it involves the identification of failure effects, modes and mechanisms. Although Fault Trees are often regarded primarily as a means of quantifying hazardous events, they are equally important as a mean of hazard identification.

Bowties

The bowtie is a way of representing potential accident dynamic, based on the coupling of a Fault Tree and an Event Tree linked through a critical event that constitutes a threat for a plant or equipment. (Chevreau et al. 2006).

Bowties saw an increase in their adoption following the EU project ARAMIS (Hourtolou & Salvi 2003) to enhance the identification of safety barriers implemented to prevent and to enhance the identification of safety barriers implemented to prevent the critical event from taking place and/or to mitigate its effects.

Safety barriers are located between sequence links. Knowing these links allows the definition of the barriers that are to be implemented in order to reduce the occurrence or the gravity of a risk represented by the critical event. Safety barriers can be technical or/and behavioural or organizational.

The strengths and weaknesses of a system exposed to an accident depend on the barrier efficiency. An accidental sequence is more likely to happen if some barriers are not operational than if all barriers are effective. Knowledge on the barrier efficiency is therefore essential for spotting the strength and weaknesses of a system. An applied example of bowtie methodology is described in Chapter 4.

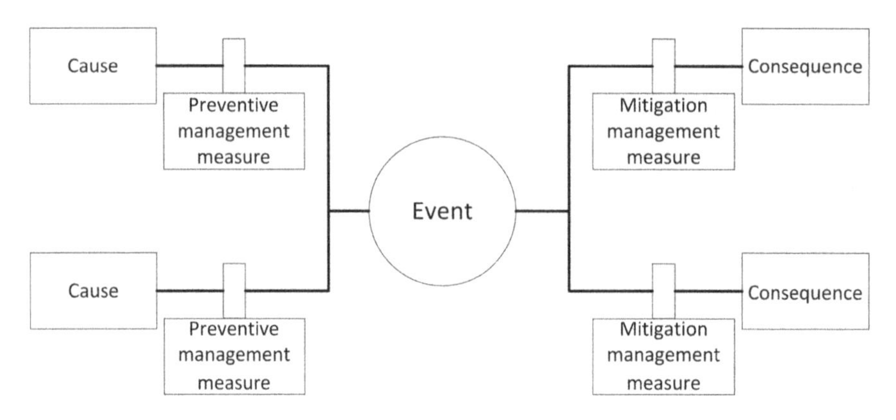

Figure 3.10 Bowtie example

Source: Cormier 2018

Estimation of consequences

The next step in the risk assessment requires an estimation of the consequences for the cases analysed with the logic trees.

The method used to estimate the consequences depends on the type of accident that is analysed and the type of expected consequence, on assets, operators, environment or the public.

In case of economic damage, it is necessary to estimate the extent of the damages caused by the accident, such as damages to the equipment, costs related to the impact on the production and any other structural damage.

The consequences on people are divided on the basis of the type of accident that is hypothesized. If a toxic release is considered, it is necessary to estimate the size of the area affected by the hazardous concentration of the substance released and the number of possible people present in an area. If a release of flammable substances is considered, the analysts should estimate the extension of the area involved by the thermal effects and the fumes, taking into account the number of possible targets involved and the consequences on them.

Thresholds to be mapped are usually defined in regulations and/or guidelines.

The description of the model or the techniques used for the estimation of the consequences of an event are out of the scope of this work.

New frontiers, dynamic risk assessment

Although the above-described methodologies are consolidated and widely accepted, they can hardly take into account time-dependent events, which are very common in process plants.

The dynamic analysis can be used in the probability assessment phase to analyse how the probability of occurrence of subsequent events is affected by the occurrence of previous events. Swaminathan and Smidts (1999) proposed the use of event sequence diagrams to allow a dynamic simulation; Bucci et al. (2008) suggested Markov modelling to build the dynamic Event Trees or Fault Trees; Čepin and Mavko (2002) proposed dynamic Fault Trees, and Khakzad, Khan, and Amyotte (2012) proposed a dynamic application of bowties.

In other cases, the dynamic analysis requires the use of a process simulator to gain insights of the evolution of the process condition in time, thus identifying critical transient conditions. A recent review of these approaches can be found in Villa et al. (2016) and Raoni et al. (2015).

More recently, thanks to the increased capabilities in on-line monitoring and data elaboration, the focus is gradually moving towards early warnings signals, deviations and past events. They can be used for risk analysis iteration, in order to progressively improve and refine the evaluated risk picture and hence reduce the probability of high impact, low probability accidents (Paltrinieri & Reniers 2017).

This section focuses on the Integrated Dynamic Decision Analysis (IDDA), which incorporates all the above-described capabilities. IDDA has been applied to risk assessment of rare procedures in major hazard installation (Gerbec et al. 2017), the analysis of plant modifications (Demichela et al. 2017) and the comparison between competing technologies for environmental protection (Baldissone et al. 2016, 2017)

This type of approach is based on a logical-probabilistic modelling of the system that is integrated with a phenomenological modelling of the system behaviour.

The logical-probabilistic model is based on general logic theory, and it is built according to the following steps, summarized in Figure 3.11:

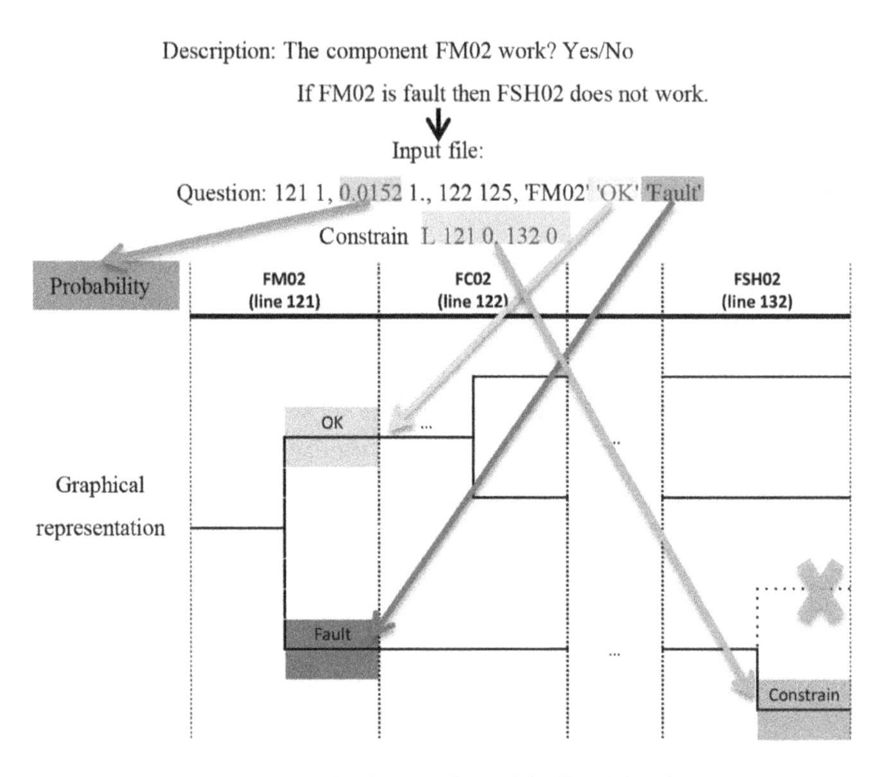

Figure 3.11 Logic Tree view of the IDDA levels

- Identification of the events related to the operation of the system, e.g. through a functional analysis, and construction of a list of levels, with questions and affirmations, which represents the elementary elements of the logical model and also the nodes in the Event Tree
- Construction of a "reticulum" indicating the addresses (subsequent level) to be visited after each response in each level, and a comment string that allows the analysts to read the logical development of a sequence
- Association to each of the levels of (i) a probability which represents the expectation degree of the failure or unwanted event and (ii) of an uncertainty ratio, which represents the distribution of the probability
- Definition of all the constraints, logical and probabilistic, which can modify the run time of the model, fitting it to the current knowledge status

Following this type of description of the system, it is possible to obtain all the possible sequences of events that the system could undergo: each sequence can describe the possible dynamic of the system with the respective probability

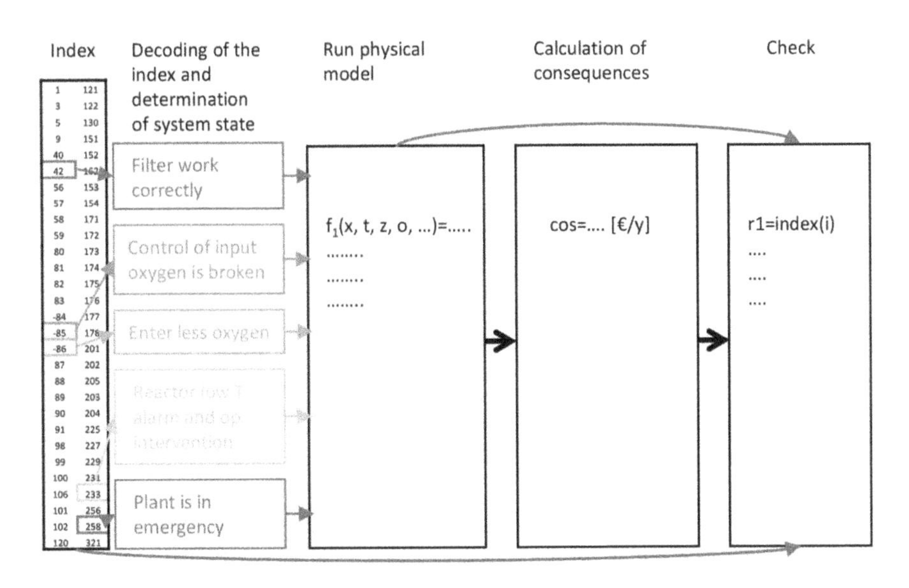

Figure 3.12 Integration of the logical-probabilistic model with the phenomenological one

Source: Baldissone 2014

The probabilistic model can run together with the phenomenological model, able to describe the physical behaviour of the system for each sequence, as summarized in Figure 3.12.

The phenomenological model of the process must be configured taking into account the system conditions described by the logical analysis to allow their communication, e.g. importing from logical modelling what equipment is expected to fail and in which way. The phenomenological model can simulate the behaviour of the system and, according to the results obtained, it may allow the logical model to update, adapting it to level of knowledge acquired by the model itself or by signals from the plant. For example, after the failure of a particular component the phenomenological model allows analysts to verify if other parts of the plant are able to compensate for its failure and the system can proceed or if cumulative effects appear and further diverge the system from its normal behaviour.

In some cases, these effects can be hardly detected during the logical analysis and the model might result not completely coherent with the reality.

This approach allows the creation of a circular pattern, represented in Figure 3.13, which implies changes to the logical analysis derived by the results of phenomenological modelling, until it adheres to reality.

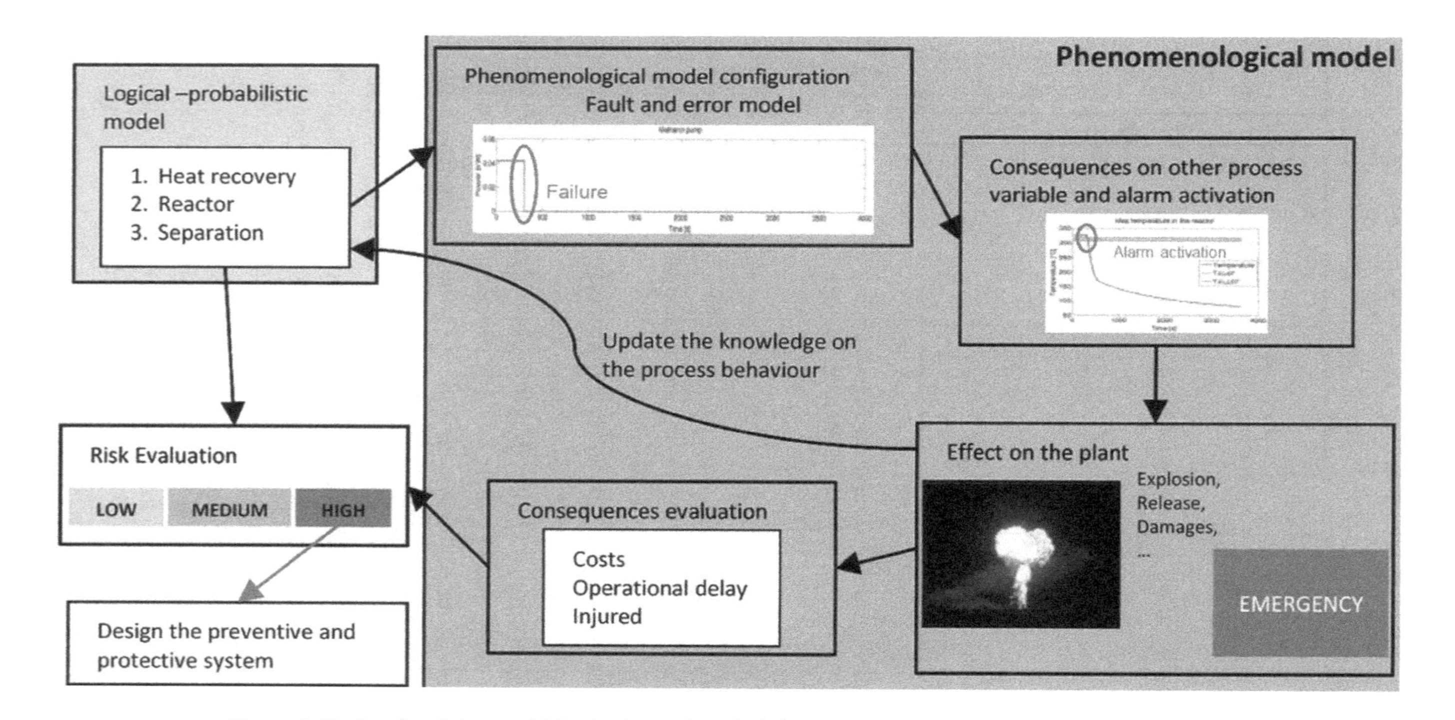

Figure 3.13 Feedback loop within the logical-probabilistic model and the phenomenological one

The phenomenological modelling can provide a direct estimation of the consequences for the various sequences in order to obtain a direct risk estimation, the evaluation of the overall risk of the system and the expected value of consequence. The latter is calculated as a weighted average of the consequences, according to their probability.

The logical modelling can be performed through a dedicated software (e.g. IDDA 2.2) which can import the consequences and get an estimation of the risk.

Within other advanced methodologies supporting risk-based decision making, the Integrated Dynamic Decision Analysis (IDDA) helps the user to decide the best design and operational solution based on a complete and consistent set of data.

With respect to traditional techniques, the integrated analysis allows to highlight the transient behaviour of the system – in case of one or more failures and/or one or more deviations – capturing the compensation and reaction capacity of the system itself.

The approach proposed allows to reduce the uncertainty both in the analysis and in the decision phase, because the results of the analysis have been based on a phenomenological model that represents the expected behaviour of the plant. Risk values (economic or not) run with the phenomenology of the process, allowing an easier comparison between the present state and the proposed alternatives.

Furthermore, the model can be quite easily updated according to the plant and process modification and can be also used to verify the possible effect of process condition modification.

Some limitations of the method have to be noticed. The quality of the results still depends on the level of detail of the model, and thus on the level of knowledge disclosed by the plant managers and risk analysts. This also affects the propagation of the uncertainties, both in the probabilistic model as well as in the phenomenological one. Moreover, the phenomenological model development can be time-consuming and complex with respect to traditional techniques, which mainly neglect it.

Conclusion and way forward

Risk assessment has been recognized as the design basis for any safety management system, as detailed in(Demichela et al. (2004 and Demichela & Piccinini (2006) and more recently reviewed in Li and Guldenmund (2018) and Pasman et al.(2017). The same view has been adopted in the TSM framework developed during the TOSCA Project.

In this view, the set of information obtained from the risk assessment should be used to the benefit of operational management of processes and plants and the risk analysis itself should take into account the information coming from the field, as well as from maintenance operation, or near misses and accident analysis, or operational feedbacks. From this virtuous loop of information exchange descends the way forward in process risk assessment.

In fact, the consolidated methodologies in process risk assessment have to be updated, and the new developed ones must be capable of exploiting the increasing amount of data that the new technologies make available to support the decision making, with a more detailed and precise view of the plant and process behaviour and response to deviations. Furthermore, thanks to the enhanced modelling capabilities, the risk assessment must be able to integrate multiple dimensions, as well as the technical and the operational and organizational ones, not as separate contributors to risk figures, but as integrated part of a complex system.

Risk assessment documentation must become a living entity, more than a static photograph of a plant or process. Also, the dynamic methodologies and the integrated systemic views, as discussed in Chapter 4, have to become the norm. These shifts of paradigm should be guided by regulations: in fact, until the policy makers would go on accepting the static risk assessment as the state-of-the-art, the companies with fewer competencies and resources will not be able to exploit the new opportunities for safer plants.

References

Baldissone, G 2014, 'Process intensification vs. reliability', PhD Thesis, Torino.

Baldissone, G, Demichela, M & Fissore, D 2017, 'Lean VOC-air mixtures catalytic treatment: Cost-benefit analysis of competing technologies', *Environments*, vol. 4, no. 46, pp. 1–18.

Baldissone, G, Fissore, D & Demichela, M 2016, 'Catalytic after-treatment of lean VOC-air streams: Process intensification vs. plant reliability', *Process Safety and Environmental Protection*, vol. 100, pp. 208–19.

Bucci, P, Kirschenbaum, J, Mangan, AL, Aldemir, T, Smith, C & Wood, T 2008, 'Construction of event-tree/fault-tree models from a Markov approach to dynamic system reliability', *Reliability Engineering & System Safety*, vol. 93, no. 11, pp. 1616–27.

Čepin, M & Mavko, B 2002, 'A dynamic fault tree', *Reliability Engineering & System Safety*, vol. 75, no. 1, pp. 83–91.

Chevreau, FR, Wybo, JL & Cauchois, D 2006, 'Organizing learning processes on risks by using the bow-tie representation', *Journal of Hazardous Materials*, vol. 130, no. 3, pp. 276–83.

Colombo, S & Demichela, M 2008, 'The systematic integration of human factors into safety analyses: An integrated engineering approach', *Reliability Engineering & System Safety*, vol. 93, no. 12, pp. 1911–21.

Cormier, R, *Bow-tie approach*, [Online] viewed 28 March 2018, www.baltspace.eu/baltspace-research/approaches-and-tools/bow-tie-approach

Demichela, M, Baldissone, G & Gianfranco, C 2017, 'Risk-based decision making for the management of change in process plants: Benefits of integrating probabilistic and phenomenological analysis', *Industrial & Engineering Chemistry Research*, vol. 56, no. 50, pp. 14873–87.

Demichela, M & Piccinini, N 2006, 'How the management aspects can affect the results of the QRA', *Journal of Loss Prevention in the Process Industries*, vol. 19, no. 1, pp. 70–7.

Demichela, M, Piccinini, N & Romano, A 2004, 'Risk analysis as a basis for safety management system', *Journal of Loss Prevention in the Process Industries*, vol. 17, no. 3, pp. 179–85.

Gerbec, M, Baldissone, G & Demichela, M 2017, 'Design of procedures for rare, new or complex processes, Part 2: Comparative risk assessment and CEA of the case study', *Safety Science*, vol. 100, no. B, pp. 203–15.

Hourtolou, D & Salvi, O 2003, 'ARAMIS Project: Development of an integrated accidental risk assessment methodology for industries in the framework of SEVESO II directive', in Bedford T, & van Gelder P (eds.), *International conference on safety and reliability (ESREL 2003)* (pp. 829–36), Maastricht.

IEC 60812 2006, *Analysis techniques for system reliability: Procedure for failure mode and effects analysis (FMEA)*, International Electrotechnical Commission, Geneva, Switzerland.

ISO-31000 2009, *Risk management*, The International Standards Organisation, Geneva, Switzerland.

Khakzad, N, Khan, F & Amyotte, P 2012, 'Dynamic risk analysis using bow-tie approach', *Reliability Engineering & System Safety*, vol. 104, pp. 36–44.

Li, Y & Guldenmund, F 2018, 'Safety management systems: A broad overview of the literature', *Safety Science*, vol. 103, pp. 94–123.

Mannan, S 2005, *Lee's loss prevention in the process industries*, Elsevier Butterworth–Heinemann, Oxford.

Paltrinieri, N & Reniers, G 2017, 'Dynamic risk analysis for Seveso sites', *Journal of Loss Prevention in the Process Industries*, vol. 49, no. A, pp. 111–19.

Pasman, HJ, Rogers, WJ & Mannan, MS 2017, 'Risk assessment: What is it worth? Shall we just do away with it, or can it do a better job? *Safety Science*, vol. 99, no. B, pp. 140–55.

Piccinini, N & Ciarambino, I 1997, 'Operability analysis devoted to the development of logic trees', *Reliability Engineering and System Safety*, vol. 55, no. 3, pp. 227–41.

Raoni, R, Secchi, AR & Demichela, M 2015, 'Employing process simulation for hazardous process deviation identification and analysis', *Safety Science*, vol. 101, pp. 209–19.

Rasmussen, NC 1975, 'Reactor safety study: An assessment of accident risks in U. S. commercial nuclear power plants: Executive summary', *WASH-1400 (NUREG-75/014): Federal Government of the United States*, U.S. Nuclear Regulatory Commission Rockville, MD, USA.

Swaminathan, S & Smidts, C 1999, 'The event sequence diagram framework for dynamic probabilistic risk assessment', *Reliability Engineering & System Safety*, vol. 63, no. 1, pp. 73–90.

Villa, V, Paltrinieri, N, Khan, F & Cozzani, V 2016, 'Towards dynamic risk analysis: A review of the risk assessment approach and its limitations in the chemical process industry', *Safety Science*, vol. 89, pp. 77–93.

Bowties for occupational risk management

Olga Aneziris and Ioannis Papazoglou

Introduction

The aim of this chapter is to present how bowties can be used to quantify occupational risks in various industrial domains. The bowtie methodology uses the Occupational Risk Model (ORM) that was developed under the auspices of the Ministry of Social Affairs and Employment in the Netherlands. Bowties are very popular in accident modelling, particularly in the process industry (Dianous & Fiévez 2006; Duijm 2009; Ferdous et al. 2012, 2013, Zuijderduijn 1999; Jacinto & Silva 2010). Their value lies mainly in their suitability for qualitative analysis. Bowties combine two well-established risk analysis methods: a fault tree (FTA), representing failures of the safety barriers that could lead to a hazard (left hand side of the model), and an event tree (ETA), representing the safety barriers put in place to mitigate the adverse consequences of the hazard (right hand side of the model).

Bowties for occupational risks are logical models that provide a way for organizing root causes into a central event that may end up in a reportable damage to the health of the worker. The uses of a bowtie model are twofold. On the one hand, it provides the accident sequences, the sequences of hazardous events that stem from fundamental or root causes and lead to the final consequences. On the other hand, it provides a way for quantifying the risk (Papazoglou & Ale 2007; Ale et al. 2008). For the purposes of quantification, the bowtie methodology has relied on two sources of data: (*a*) detailed descriptions of work-related accidents that occurred in the Netherlands over a certain period of time and were investigated by the Dutch Labour Inspectorate of the Ministry of Social Affairs and Employment and (*b*) data on exposure assessment (i.e., the time for which the working staff were exposed to the hazards).

In addition, the frequencies of working conditions in the workplace underlying the onset of accidents were assessed. This second source of data was generated through surveys of the Dutch working population at a national level. Data for the development of these models were derived from the GISAI database (GISAI 2005) of the Netherlands Ministry of Work, which incorporates 23,740 occupational accidents that occurred between 1998 and 2011. A case study will be presented in paragraph "Bowtie model for in or on moving vehicles, with loss of control" to analyse activities performed near vehicles in instances where workers are in or on a moving vehicle that loses control. In total 63 bowties have been developed for all 23,740 occupational accidents which may occur in the working population. The details of these bowties are given in RIVM (2008).

Bowties and occupational risk

Occupational risks from accidents are quantified when we can measure the severity of the potential health damage and can assess how likely each level of damage is expected to occur. Every time an individual performs a work-related activity and faces one or more hazards they face the possibility of an accident that may result in bodily harm. Such a consequence does not happen with certainty, but rather occasionally, in the working population and during the lifetime of the workers. The associated risk can be quantified when we know the probabilities that different consequences can occur during a specified period of time. In this model of harm, each worker can suffer three possible consequences:

- recoverable injury
- permanent injury
- death

This distinction is related to the Dutch occupational accident reporting requirements for serious accidents that address fatal, permanent injury and hospital treatment outcomes (GISAI 2005).

Building the bowtie model for a single-hazard

The basis of ORM is the model for a single-hazard. For a specific accident type (e.g., fall from height) the ORM model provides a framework for

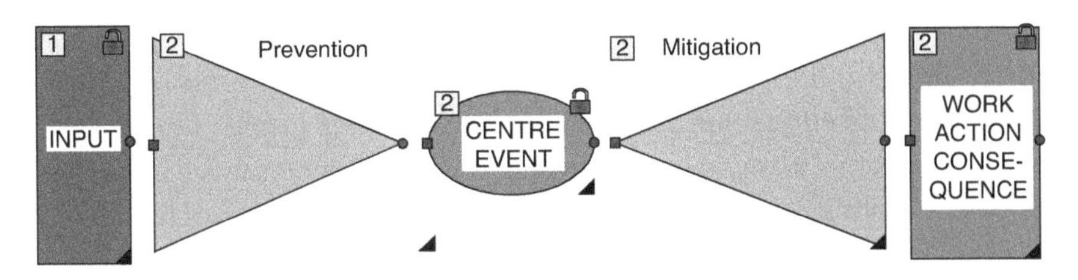

Figure 4.1 The bowtie model

making a quantified link between exposure to the causes and effect factors and their relation to the potential outcomes. The single-hazard ORM model has been called a "bowtie" (see Figure 4.1). The concept of the "bowtie" is based on the definition of an important hazardous event (or hazard) that describes the main characteristic of an accident. Such hazardous events are, for example, a "fall from height" or a "contact of part of the body with a moving part of a machine". Appropriately, this event is called the "Centre Event" (CE). Then the model can be distinguished in two parts: left of the "centre event" called the left hand side (LHS); and right of the "centre event" called the right hand Side (RHS).

The LHS of the bowtie depicts the part of the model that includes the causes that lead to the centre event. In this part of the model, the accident evolves in time going from left to right. Once the centre event has occurred, the bowtie model develops on the right-hand-side. This part depicts all possible and consecutive events leading to the final consequence. In principle, the LHS of a bowtie includes all the elements of the accident sequences that correspond to "prevention" measures: whatever means exist that aim at preventing the centre event from occurring. Similarly, the RHS of a bowtie includes all elements that influence the final consequence given that the centre event has occurred and hence it represents s all the "mitigation" aspects of an accident.

A fundamental concept of the "bowtie" model is the concept of a "safety barrier". A safety barrier is a physical entity, a technical, hardware, procedural or organizational element in the working environment that can prevent something from happening (e.g. the CE) or mitigate the consequences of something that has happened. Some of these events or barriers have a deterministic logical relationship with the event that follows them or the event that they cause. For example, "loss of a worker's stability while on a high platform" coupled with the "absence or failure of an edge protection-rail" may result in a fall from height. On the other hand, a second class of events in the model influences the relative likelihood of other events occurring; they do not determine the outcome of more complex events. Thus, "working outdoors" in a "windy environment" does not necessarily imply "loss of worker stability" but it rather increases the likelihood that such an event might happen.

The elements of this logic model are the Centre Event, the primary safety blocks, the support safety blocks, as presented in Figure 4.2 for Left Hand Side of the bowtie with a Centre event "struck by moving vehicle". In addition Consequences are required in order to estimate risk.

A) **Centre Event**: this event that determines whether a specific hazard (e.g. struck by moving vehicle) has occurred or not. This event has two states.
   CE(+): The specific hazard has not occurred
   CE(−): The specific hazard has occurred
B) **Primary Safety Blocks**: their outcomes determine whether an accident with reportable Consequence has occurred or not. A Primary Safety Block (PSB) models a technical element, a hardware element, a physical condition or a property in the working environment that aims either at preventing an accident from happening or at mitigating its consequences. For this reason, they are sometimes called "safety barriers". Each PSB models an event that can have one out of a number of distinct, discrete and mutually exclusive outcomes. Equivalently a PSB can be considered as a component that can be in one of a number of distinct, discrete and mutually exclusive states. If there are z PSBs relevant to the prevention of a hazard then these PSBs form a system that determines whether the hazard has occurred or not.

   Knowledge of the probabilities of occurrence of each system state provides a probabilistic linkage of the PSBs and the probability of occurrence of an accident. If each and every one of the PSBs can be linked to a single, simple, easily understood workplace condition, or other associated factor that can be affected by an equally specific action, then the model development stops here. This is not always the case, however, since in most cases PSBs are not that simple. In these cases, the model is further developed by the introduction of the Support Safety Blocks.
C) **Support Safety Blocks**: model events whose outcomes influence the probability with which the states of the PBSs are obtained. A Support Safety Block (SSB) models a physical entity, a technical, hardware,

procedural or organizational element in the working environment that aims at supporting the success of the primary safety block for this reason they are sometimes called "Support Safety Barriers".

Each SSB can be in two states:

State (+): Support Safety Block is successful, good, proper or adequate.

State (-): Support Safety Block is unsuccessful, bad, improper or inadequate.

D) **Consequences:** this event determines whether an accident with reportable consequences has occurred or not. This event has four states.

c1: No accident or an accident with non-reportable consequences (considered minor).

c2: An accident resulting in a reportable recoverable injury.

c3: An accident resulting in a permanent injury.

c4: An accident resulting in a fatal injury.

Consequences belong to the Right Hand Side of a bowtie, addressing events and their outcomes, which in combination with the outcomes of the centre event determine the overall consequences.

Figure 4.2 presents the LHS of the bowtie with central event "struck by moving vehicle" which has one Primary Safety Block, "Vehicle moving towards a pedestrian", and eight support safety barriers that influence the probability of the PSB. The eight SSBs are as follows:

a) Lock-out safety system. A vehicle should be provided with a lock-out safety system so as to prevent a sudden movement of the vehicle.
b) Mechanical integrity of vehicle. A vehicle should be in good mechanical condition.
c) Vehicle velocity. A vehicle should not perform over speeding.
d) Driver ability. The driver should have good driving skills, no physical limitations and behave responsibly.
e) Visual contact of driver. The driver should have full view of pedestrian routes.
f) Visual/audible contact of pedestrian. The pedestrian should have full view of vehicle route.
g) Ability of pedestrian. The pedestrian should be able to participate in the traffic.
h) Location/position of pedestrian. The pedestrian should be in proper location/position

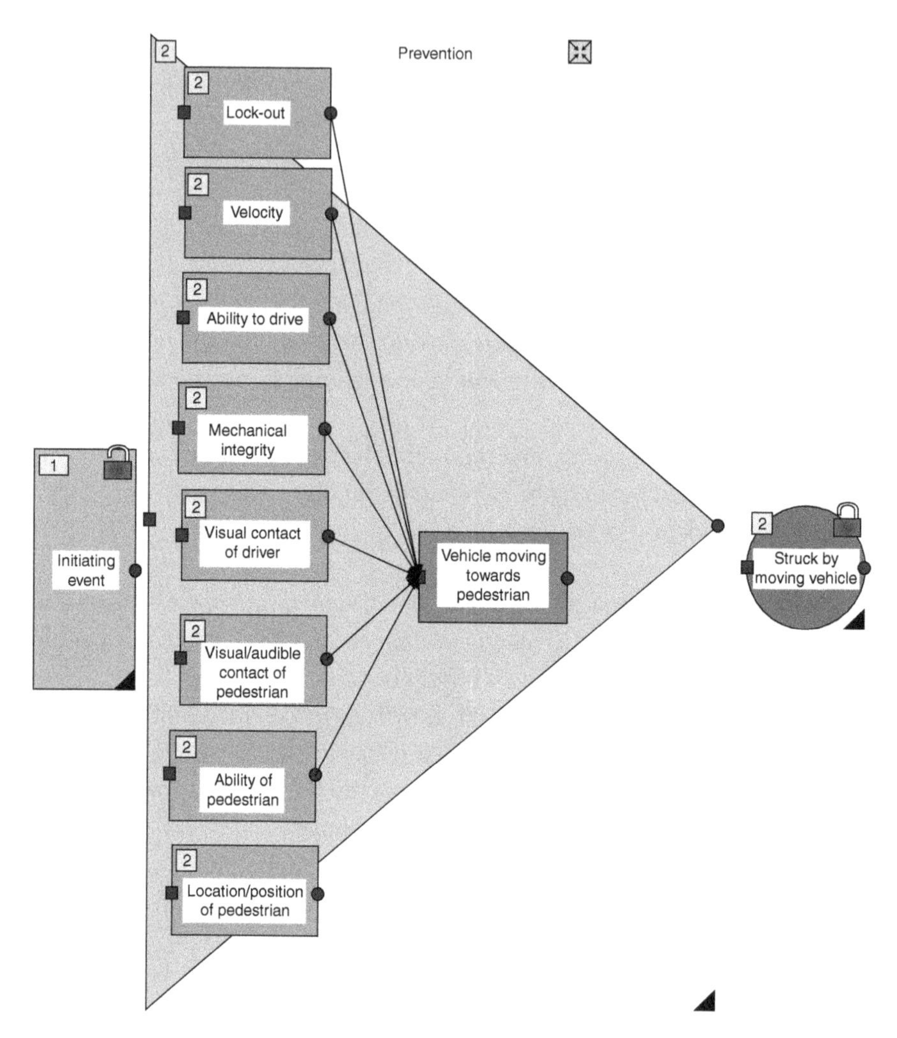

Figure 4.2 Barriers and Centre event of a bowtie for "struck by moving vehicle"

The development of a bowtie consists of the following steps:

i) Determination of the Primary Safety Blocks that characterize the occurrence of an accident with a reportable consequence.
ii) Determination of the Support Safety Blocks expressing those aspects of the working conditions that influence the success or failure of the PSBs.
iii) Assessment of the influences among the PSBs and the SSBs in order to determine blocks that are influenced by other blocks in the diagram.

iv) Determination of the specific consequence of each and every of the possible state of the combined set of PSBs and SSBs. Papazoglou et al. (2007) provide all the details for the development of this single-hazard logic model of bowtie.

In the ORM project, 63 specific single-hazard models have been developed and presented in the RIVM report and by Ale (2006) and by Papazoglou et al. (2017). Building a logical model for a single-hazard is based on the existing characteristics of the working environment, the fundamental laws of physics and engineering as well as the characteristics of human behaviour. The models and their quantification are based on the observed historical accident data in the Netherlands. These include the accident reports of the Dutch Labour Inspectorate in their management database GISAI (2005). This information has been analysed and organized in a systematic way that is consistent with the adopted logic on the single-hazard model. In particular, a method and an associated computer tool named "Storybuilder" have been developed by Bellamy et al. (2007). All the "structural" information – i.e. safety barriers, their failure modes, combinations of failures leading to more general events, and eventually through potentially mitigating effects to the final consequence – are contained in the storybuilds developed from the accident reports that detail the causes and effects of the particular hazards (e.g., struck by moving vehicle).

Quantification of the bowtie

The Occupational Risk Model needs the following quantitative data for risk calculations:

1 unconditional probabilities of SBBs and/or PSBs and conditional probabilities for blocks which receive an influence from others (e.g. conditional probabilities of primary barriers given the state of support barriers)
2 exposure of workers to the single-hazard

These probabilities were obtained from two sets of data:

(i) unconditional probabilities of support safety blocks and some primary safety blocks obtained through surveys and assessment of working conditions in the Dutch working population

(ii) accident sequence probabilities obtained from observed accident statistics and assessments of the exposure of the working population that was engaged in these accidents

In several instances, the blocks in the bowties were simple enough so that their probability of occurrence could be directly assessed through the survey. In other instances, however, this was not possible. Some of the low-level support barriers required additional decomposition to even simpler events that somehow collectively determined the outcome of the decomposed events. For example, consider the following two events (SSBs) among the various underlying causes for "a struck by vehicle" accident: (*a*) "Mechanical integrity (of vehicle)" and (*b*) "Ability to drive" (see Figure 4.2). Assessing the frequency with which the mechanical conditions of a vehicle are substandard is straightforward since the meaning of a failure of mechanical integrity is easily understood. Assessing the frequency of the ability to drive is not that straightforward however, since "ability to drive" includes many factors. For this reason, this event had to be decomposed into other more detailed and more concrete events or such as: (*a*) driver's license (*b*) physical state of driver and (*c*) alertness of driver. These factors are called "Probability Influencing Entities" (PIEs).

Each influencing factor (PIE) is assumed to have two possible levels: "Adequate" and "Inadequate". The quality of an influencing factor is then set equal to the frequency with which this factor is at the adequate level in the working places. Then the quality of the barrier is given by a weighted sum of the influencing factor qualities. The weights reflect the relative importance of each factor and are assessed by the analyst on the basis of expert judgement. Currently, equal weights have been used. This way, the probability that a support barrier is in one of its possible states is given by the weighted sum of the frequencies of the influencing factors (see Papazoglou et al. 2017; RIVM 2008).

The probabilities of consequences (i.e. fatality, permanent injury and recoverable injury) have been quantified as the ratio of the number of instances that a particular event outcome occurred and the observed exposure. The denominator refers to the total number of instances the event has been observed or the total duration over which the occurrences have been observed.

Risk reducing measures

Usually, safety managers propose specific measures with the intention to improve occupational safety. The most efficient way to evaluate the effectiveness of a particular measure is to assess its effect on the risk of an occupational accident. The proposed model offers a direct way to measure the risk reduction implied by a specific measure.

To reduce the risk, measures can be introduced to affect the state of the PIEs. A measure is defined as a collection of specific actions that result in a change of the probabilities of PIEs. These changes are then transferred into a corresponding change in the quality of the safety barrier and this in turn to the probabilities of the remaining PSBs in the single-hazard model. Hence, changes in the probabilities of the consequences result in a measurable change in the risk. In the next sections, several risk reduction measures are discussed for the bowtie of the Central Event "in or on moving vehicles, with loss of control".

Bowtie model for "in or on moving vehicles, with loss of control"

This section presents a logical model for the "in or on moving vehicle, with loss control" event and it has been based on 541 accidents that occurred in the Netherlands as well as on information about safety rules concerning activities performed while working near vehicles.

Left Hand Side of the bowtie (LHS)

The Left Hand Side of the bowtie consists of the initiating event and corresponding safety measures that could prevent "loss of control" of a vehicle with workers in or on it (Figure 4.3). The initiating event represents the following situations: (*a*) the worker is the driver of a moving vehicle, (*b*) the worker is a passenger of a moving vehicle which may be driving ahead, in reverse, manoeuvring or standing still, and (*c*) the worker is handling load(s), repairing, inspecting (or other specific activities with or related to the vehicle).

An analysis of 541 real accidents has determined four ways in which a worker might get injured due to loss of control of the vehicle, as follows: (*a*) the worker loses his stability while he is in or on a moving vehicle; (*b*) the worker inside the vehicle hits an object which may be outside the vehicle; (*c*) the vehicle loses its stability and (*d*) the vehicle turns over or falls down over an edge. Correspondingly, there are four primary safety functions that could prevent these failure modes. The bowtie includes four Primary Safety blocks namely: "User

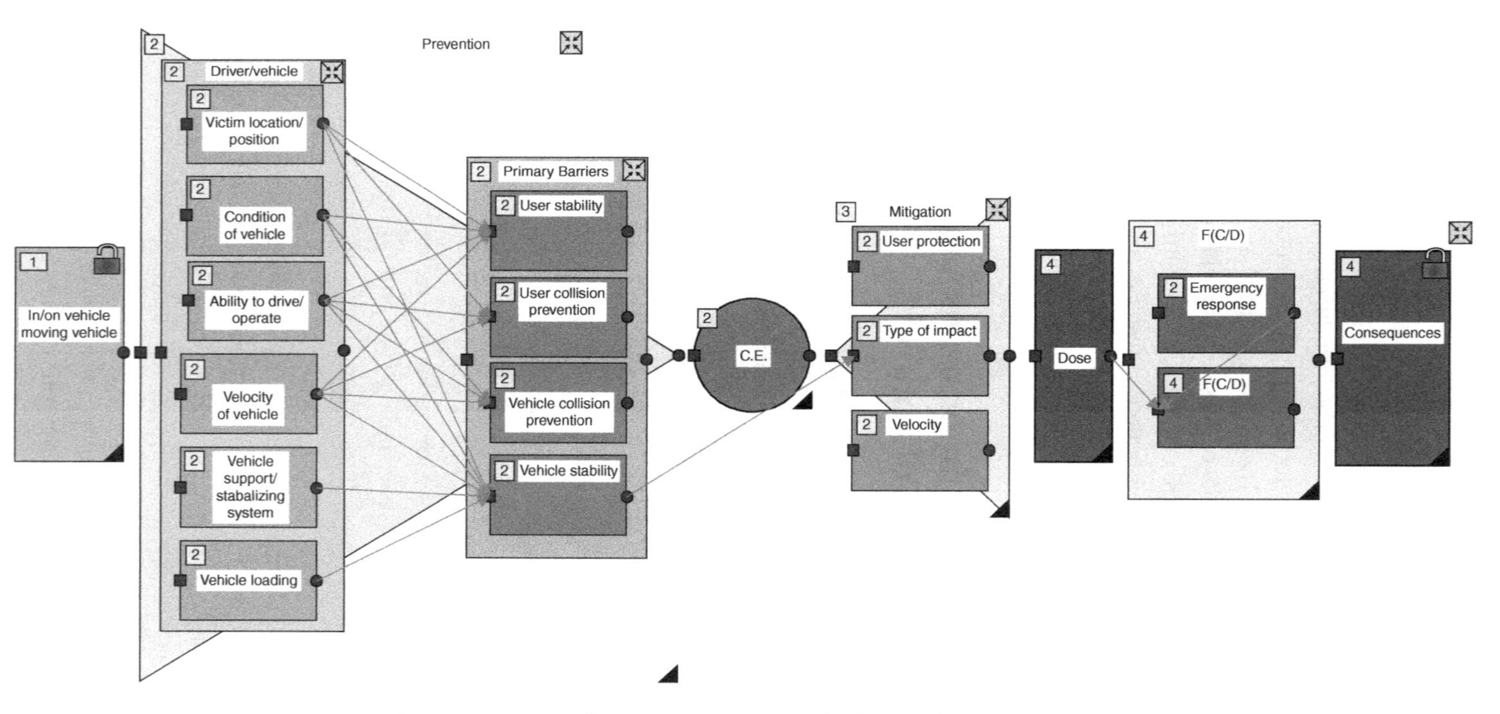

Figure 4.3 Bowtie for "in or on moving vehicle with loss of control"

Stability", "User Collision Prevention", "Vehicle Collision Prevention", and "Vehicle Stability". Primary Safety Blocks may be in two states, for example:

PSB1: User Stability may be in one of the following two states
PSB1(+): User in balance
PSB1(–): Loss of user balance

Support Safety Barriers (SSBs) contribute to the adequate function of the Primary Safety Barriers and influence the probability with which the primary safety barrier-states occur. In this bowtie, there are seven support barriers, namely:

a) Victim location/position: The driver or passenger of vehicle should be situated at a safe place
b) Condition of vehicle: The vehicle should be in a good technical state (mechanical, electrical and structural)
c) Driver ability
d) Vehicle velocity
e) Vehicle support/stabilizing system: The vehicle should have good technical means for keeping it stable during use
f) Vehicle loading: The vehicle should be properly loaded
g) User protection: Safety belts should be used and one or more protection provisions should be present

More details concerning the support barriers of both bowties are presented by Baksteen et al. (2008).

Right Hand Side (RHS)

Events in the Right Hand Side of the bowtie are combined with the outcome of the Centre Event and determine the consequences of the contact. Four levels of consequences are used: C1: No consequence; C2: Recoverable injury; C3: Permanent injury; C4: Death. Events of the RHS are the type of contact between vehicle and worker, the velocity of the vehicle and the emergency response.

Risk quantification

Individual risk of death, permanent injury and recoverable injury per hour have been assessed for the bowties, according to the methodology presented by Aneziris et al. (2008) and Papazoglou et al. (2008, 2015). Quantification of the model given a yearly exposure of 4.31×10^8 hours of the Dutch working population for bowtie "In or on moving vehicle with loss of control" resulted in the following probabilities for an hour of exposure to this hazard:

Probability of loss of control: 2.00×10^{-7}/hr
Probability of Recoverable Injury: 1.25×10^{-7}/hr
Probability of Permanent Injury: 6.14×10^{-8} /hr
Probability of Fatality: 1.394×10^{-8} /hr

Failure probabilities for the barriers, PIEs and their frequencies are presented in Table 4.1. Frequencies of PIEs have been assessed through surveys of the working condition in the Dutch working population and reflect the Dutch National Average (RIVM 2008). For example, the event "Victim location/position" has been decomposed into other more detailed and more concrete factors such as: (*a*) premature exit/ entrance of vehicle by passenger, (*b*) position of body parts outside the vehicle, (*c*) position of person in or on vehicle and (*d*) location of vehicle. The failure probability of these events, as assessed, are presented in third column of Table 4.1 and the Failure probability of the barrier "Victim location/position" is equal to 0.0925 as presented in the fourth column of Table 4.1.

Importance analysis

To assess the relative importance of each factor influencing this particular risk we have calculated two importance measures:

1) Risk decrease with respect to the present state, if a barrier (or PIE) achieves its perfect state with probability equal to unity.
2) Risk increase with respect to the present state, if the barrier (or PIE) achieves its failed state with probability equal to unity.

Risk decrease prioritizes the various elements of the model for the purposes of possible improvements. It is more effective to start improving firstly a barrier with a high risk decrease potential than another with a low risk decrease potential.

Risk increase potential provides a measure of the importance of each element in the model to be maintained at its present level of quality. It is more important to concentrate on the maintenance of a barrier with a high risk

Table 4.1 PIES Characteristics and frequencies for bowtie "in or on moving vehicle with loss of control"

Primary Barriers	PIE	PIE Frequency	Barrier Failure Probability
Victim location/ position	Premature exit/entrance of vehicle by passenger	0.06	0.0925
	Position of body parts outside the vehicle	0.09	
	Position of person in or on vehicle	0.1	
	Location of vehicle	0.12	
Condition of vehicle	Mechanical state of vehicle	0.07	0.07
Ability to drive/ operate	No license to drive	0.07	0.105
	Physical state of driver	0.06	
	Alertness of driver	0.18	
	State of surface	0.16	
	Marking/ Signalling	0.08	
	Unintentional movement	0.06	
	Obstruction obstacles	0.19	
	Substandard visual conditions	0.11	
	Premature exit of vehicle by driver	0.04	
	Lay-out/profiel van de weg	0.1	
Velocity of vehicle	Safe speed	0.18	0.18
Vehicle support/ stabilizing system	Bad weather condition – High wind speeds/Squalls	0.09	0.055
	Loading platform	0.06	
	Vehicle over edge and placement failure	0.06	
	counter weight and stabilizing-legs failure	0.01	
Vehicle loading	Load or load connection failure	0.06	0.065
	Fork/load too high when driving	0.07	
User protection	Safety belt	0.28	0.217
	PPE	0.23	
	Vehicle impact protection	0.14	

increase importance than one with a lesser one. The effect each PIE has on the overall risk is presented in Table 4.2.

"Driving a vehicle safely within safe speed limits" is the most important barrier that decreases fatality risk with high indexes for the three consequences levels (Decrease fatality rate is 59%). "Good mechanical state of the vehicle" (i.e., braking on time and moving/steering properly) is the most important barrier to maintain fatality risk level with index 616%.

Table 4.2 Fatality risk importance measures for each PIE for "in or on moving vehicle with loss of control"

Safety barrier	Risk Decrease rate (/hr)	Risk Decrease	Risk increase rate (/hr)	Risk Increase
Base case	1.39E-08			
Premature vehicle exit/ entrance by passenger	1.27E-08	8.74E-02	3.27E-08	1.35
Position of body parts outside the vehicle	1.21E-08	1.31E-01	3.21E-08	1.31
Position of person in or on vehicle	1.19E-08	1.45E-01	3.19E-08	1.30
Vehicle location	1.15E-08	1.74E-01	3.15E-08	1.27
Mechanical state of vehicle	7.45E-09	4.64E-01	1.09E-07	6.16
No driving license	1.31E-08	5.42E-02	2.37E-08	7.07E-01
Physical state of driver	1.33E-08	4.66E-02	2.38E-08	7.14E-01
Drivers alertness	1.2E-08	1.38E-01	2.26E-08	6.23E-01
State of surface	1.22E-08	1.23E-01	2.28E-08	6.38E-01
Marking/Signalling	1.3E-08	6.18E-02	2.36E-08	6.99E-01
Unintentional movement	1.33E-08	4.66E-02	2.38E-08	7.14E-01
Obstruction obstacles	1.19E-08	1.45E-01	2.25E-08	6.15E-01
Substandard visual conditions	1.27E-08	8.46E-02	2.33E-08	6.76E-01
Premature drivers vehicle exit	1.35E-08	3.14E-02	2.4E-08	7.29E-01
Lay-out	1.28E-08	7.70E-02	2.34E-08	6.84E-01
Safe speed	5.74E-09	5.87E-01	5.1E-08	2.67
Bad weather –	1.24E-08	1.08E-01	2.89E-08	1.08
Loading platform	1.29E-08	7.20E-02	2.94E-08	1.11
Vehicle over edge and placement failure	1.29E-08	7.20E-02	2.94E-08	1.11

Counter weight and stabilizing-legs failure	1.37E-08	1.28E-02	3.02E-08	1.17
Load or load connection failure	1.29E-08	7.46E-02	3.15E-08	1.27
Fork/load too high when driving	1.29E-08	7.46E-02	3.15E-08	1.27
Safety belt	1.01E-08	2.75E-01	2.37E-08	7.05E-01
PPE	1.08E-08	2.26E-01	2.44E-08	7.54E-01
Vehicle impact protection	1.20E-08	1.38E-01	2.56E-08	8.42E-01

Conclusions

Occupational risk assessment is a major part in Total Safety Management of installations as workers are exposed to many hazards in their daily working conditions. Quantification of occupational hazards can be performed by the bowtie method that was presented here. The bowtie model consists of three levels of events: (*a*) the Primary Safety Barriers (PSB) representing the immediate causes of the Centre Event (CE); (*b*) the Secondary Safety Barriers (SSB) representing events that influence the probability of the PBSs and (*c*) the Probability Influence Entities (PIE) that determine the probability of the SSBs. The last level of developed events (the PIEs) consists of events simple enough that their expected frequency can be quantified much easier than that of the complex composite event. Furthermore, their expected frequency can be influenced by simple straightforward actions. As a case study the risk assessment of the bowtie "in or on a vehicle which loses its control" has been presented and risk reducing measures have been proposed and assessed. The most important measure for fatality risk reduction is safe speed, followed by the good mechanical state of the vehicle.

References

Ale, BJM 2006, 'The occupational risk model', TU-Delft/TBM RC 20060731, ISBN 90-5638-157-1, Delft.

Ale, BJM, Baksteen, H, Bellamy, LJ, Bloemhof, A, Goossens, L, Hale, AR, Mud, ML, Oh, JIH, Papazoglou, IA, Post, J & Whiston, JY 2008, 'Quantifying occupational risk: The development of an occupational risk model', *Safety Science*, vol. 46, no. 2, pp. 176–85.

Aneziris, ON, Papazoglou, IA, Baksteen, H, Mud, ML, Ale, BJM, Bellamy, LJ, Hale, AR, Bloemhoff, A, Post, J & Oh, JIH 2008, 'Quantified risk assessment for fall from height', *Safety Science*, vol. 46, no. 2, pp. 198–220.

Baksteen, H, Mud, M, Bellamy, L, Papazoglou, IA, Aneziris, O & Konstantinidou, M 2008, 'Technical report on the modelling of Bowties 11: In or on moving vehicle with loss of control', *WORM Metamorphosis Report*, Bilthoven, The Netherlands.

Bellamy, LJ, Ale, BJM, Geyer, TAW, Goossens, LHJ, Hale, AR, Oh, JIH, Mud, ML, Bloemhoff, A, Papazoglou, IA & Whiston, JY 2007, 'Storybuilder: A tool for the analysis of accident reports', *Reliability Engineering and System Safety*, vol. 92, pp. 735–44.

Dianous, V & Fiévez, C 2006, 'ARAMIS project: A more explicit demonstration of risk control through the use of bow: Tie diagrams and the evaluation of safety barrier performance', *Journal of Hazardous Materials*, vol. 130, no. 3, pp. 220–33.

Duijm, NJ 2009, 'Safety-barrier diagrams as a safety management tool', Reliability *Engineering & Systems Safety*, vol. 94 no. 2, pp. 332–41.

Ferdous, R, Khan, F, Sadiq, R, Amyotte, P & Veitch, B 2012, 'Handling and updating uncertain information in bow-tie analysis', *Journal of Loss Prevention in the Process Industrie*s, vol. 25, no. 1, pp. 8–19.

Ferdous, R, Khan, F, Sadiq, R, Amyotte, P & Veitch, B 2013, 'Analyzing system safety and risks under uncertainty using a bow-tie diagram: An innovative approach', *Process Saf. Environ. Prot.*, vol. 91 nos. 1–2, pp. 1–18.

GISAI 2005, 'Geintegreerd informatie systeem arbeids inspectie: Integrated information system of the labor inspection in the Netherlands', Den Haag, The Netherlands.

Jacinto, C & Silva, C 2010, 'A semi-quantitative assessment of occupational risks using bowtie representation', *Safety Science.*, vol. 48, pp. 973–9.

Papazoglou, IA & Ale, BJM 2007, 'A logical model for quantification of occupational risk', *Reliability Engineering & System Safety*, vol. 92, no. 6, pp. 785–803.

Papazoglou, IA, Aneziris, ON, Bellamy, LJ, Ale, BJM & Oh, J 2017, 'Quantitative occupational risk model: Single hazard', *Reliability Engineering and System Safety*, vol. 160, pp. 162–73.

Papazoglou, IA, Aneziris, ON, Bellamy, LJ, Damen, M, Ale, BJ & Oh, JIH 2015, 'Uncertainty assessment in the quantification of risk rates of occupational accidents', *Risk Analysis*, vol. 8, pp. 1536–61.

Papazoglou, IA, Bellamy, LJ, Leidelmeijer, KCM, Damen, M, Bloemhoff, A, Kuiper, J, Ale, BJM & Oh, JIH 2008, 'Quantification of occupational risk from accidents', *PSAM*, vol. 9.

RIVM 2008, 'WORM metamorphosis consortium: The quantification of occupational risk: The development of a risk assessment model and software', *RIVM Report 620801001/2007*, The Hague.

Zuijderduijn, C 1999, 'Risk management at Shell Pernis refinery/chemicals: Implementation of SEVESO-II based on build up experiences, using a hazards and effects management process', in *Proceedings of the Seveso 2000 European conference*, Athens, Greece, 10–12 November.

A system dynamics approach in modeling business processes in safety management and production

Tom Kontogiannis, Dimitris Boukas and Christina Malaki

Introduction

One of the challenges of Total Safety Management (TSM) is the modeling of interactions of business processes involved in safety, quality and production. Such business processes may include the way that tasks are organized (i.e., task management), the patterns of work and training of workers (i.e., human resources management), the management of errors and rework and so forth. It is foreseeable that interventions in safety may adversely affect other organizational interests in production, finance and quality. In other cases, safety investments may take a long time to produce tangible benefits which could result in losing managerial enthusiasm. Understanding and modeling how changes in safety management affect the organization as a whole could provide a good basis for making better decisions about safety investments.

System dynamics have been used in the past (Sterman 2000; Morecroft 2007) for modeling the interactions of business processes in finance, production and quality control. In the last few years, some studies have also looked into the interactions between safety and production using quantitative approaches (Cooke 2003; Lyneis & Madnick 2009; Bouloiz et al. 2013). These studies have some advantages in managing risks as they focus on the analysis of the organizational control structure, the safety culture, the interactions of technical and organizational factors as well as the modeling of the dynamics of the system. In this approach, accidents are not seen as the immediate result of a disturbing factor or an event but rather the cumulative effect of a number of interacting factors with their own dynamics.

Within the system theoretic approach (Leveson 2011), system dynamics have been used in modeling organizational trade-offs between production and safety (Marais et al. 2006; Goh et al. 2012; Kontogiannis 2012). For instance, an emphasis on production may be at the expense of safety and result in a higher incident rate because task management could be inadequate. In turn, the higher cost of incidents may have repercussions for the financial resources and, hence, may result in a lower emphasis on production. Similarly, a higher incident rate may increase the attrition rate of workers, as they would feel that there may be a high risk for their lives. Some industries (e.g., mining) have a relatively high rate of turnover which creates a training problem for the organization and hence, productivity may be reduced in the short term. In this sense, it is possible to build a model of the overall system that brings together the interactions between safety, quality and production.

This chapter aims to explore the use of systems dynamics in the context of TSM in order to examine the interactions of safety with other organizational domains as well as look into the time dynamics of Key Performance Indicators (KPIs) in the near future. The TSM approach to safety provides a good basis for looking deeper into the interaction of technical, workplace and organizational factors, their magnitude of effects and their time characteristics. It focuses on the organizational processes that control different goals (e.g., safety, human resources, production) and the trade-offs that should be managed in the context of the overall system.

A case study of a typical mine

A case study of a typical mine is utilized in this chapter which had a production rate of 5,000 tons/week and employed a number of 55 miners who would perform 250 tasks/week. Miners had an average productivity of 90 tons/week/person and would work at shifts of 40 hours/week. An index of Production Pressure was calculated for management as the ratio of production rate and customer orders. It was expected that as Production Pressure increases, management would be more likely to get absorbed in production matters and lose sight of safety. On the other side, workers were influenced by the Schedule Pressure to get their tasks ready in time and minimize task backlog. As a result, an index of Schedule Pressure was used that corresponded to an index of workload of workers which could deteriorate their ability to recover errors.

In principle, the proposed model can be applied to a variety of industries provided that an extension is made to the ways in which increases in production are managed. For instance, investments in new technology play an important role in managing production challenges. In this case study, the focus was mainly on the adjustment of workforce and worker productivity in managing production issues.

A system dynamics model of safety and production

System dynamics represent systems as causal loop diagrams showing the control loops and influences stemming from internal and external variables. There are two basic types of loops which combine in complex ways, interact in dynamic ways and produce effects that are difficult to foresee in the first place. A Balancing loop (or a negative loop) is a structure that bridges the gap between an actual and desired state through some action. It is a typical control structure that maintains an important variable close to a target point which is achieved after a period of fluctuation around it. System changes that upset a variable of interest (e.g., incident rate) take time to produce effects, which creates a transitional period where the variable fluctuates around its final value. A Reinforcing loop (or positive loop) usually generates exponential changes of variables and reinforces their current trend of growth or decay. It can be considered either as a positive function that reinforces a positive trend or as a negative function that creates side effects to a balancing loop.

In general, the interactions between production and safety take place within a wider organizational context that also includes other control structures such as human resources management and task management. Adverse interactions are manifested as incidents, unsafe conditions, losses of production and departure of experience workers. The control structures of four subsystems and their interactions can be seen in the following causal loop diagram (Figure 5.1), which provides a high level overview of the causal structure; the sections that follow describe individual modules of the dynamic model of the system.

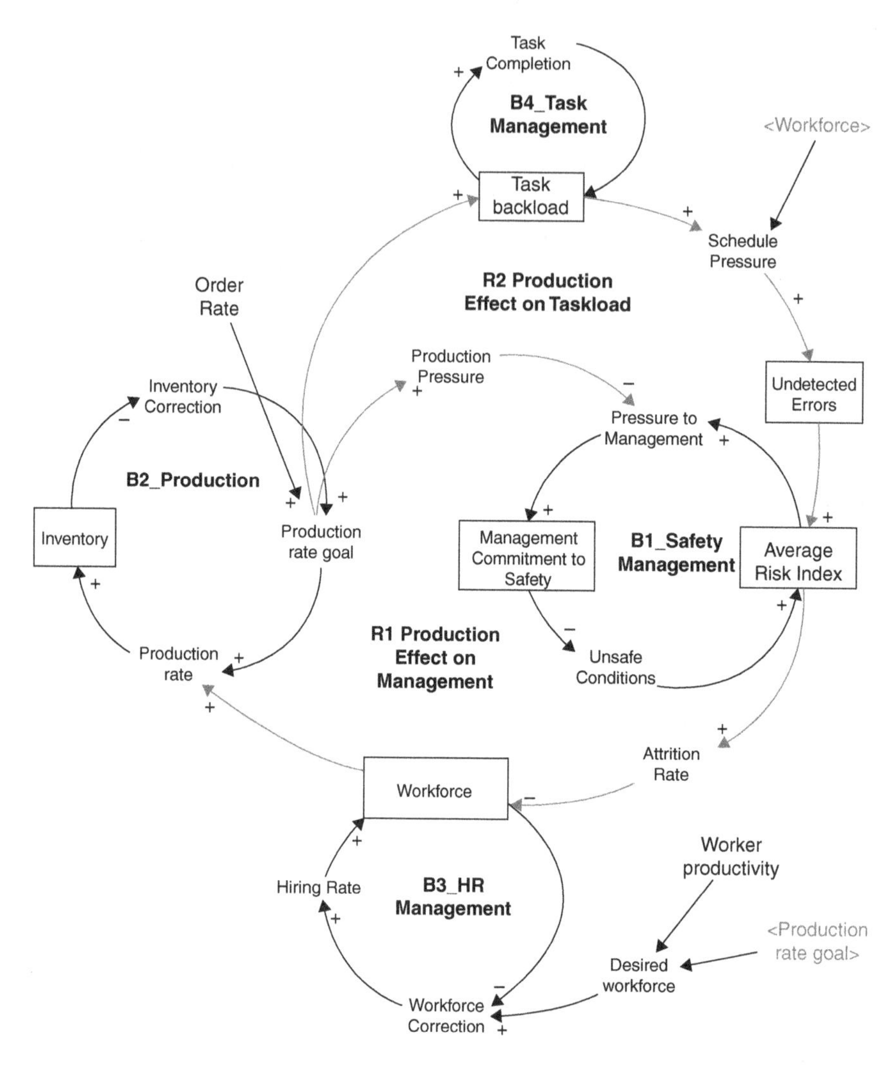

Figure 5.1 Overview of causal loops of four subsystems in coal mining

For simplicity reasons, only the most prominent control loops are protrayed in Figure 5.1. The arrows between variables indicate that they may change either in the same direction (i.e., a plus sign) or in a different direction (i.e., a minus sign). As it can be seen, each subsystem is mainly controlled by a balancing loop (B) that brings stability after a period of disturbance in the variables of the model. The emphasis of our study has been more on the regulation and interaction between safety and production requirements. On the one hand, production can influence safety through Production Pressure from orders and Schedule Pressure from tasks. Specifically, Production Pressure can lower Management Commitment to Safety (MCS) and increase unsafe conditions while Schedule Pressure can increase Task Backlog and the number of Undetected Errors (Figure 5.1).

On the other hand, safety could affect production through the human resources system. In specific, a higher incident or risk index rate could increase the attrition rate of workers and reduce the level of workforce and, hence, the inventory of orders. The interactions between safety and production can create two reinforcing or vicious loops where increases in production could result either in lower Management Commitment to Safety or higher Undetected Errors. In both cases, the incident rate or risk index would increase, which, in turn, could lower inventory of orders and, hence, further increase the Production Rate Goal. Unfortunately, these production reinforcing loops tend to create an increase of the incident rate or risk index.

Safety system

The modeling of safety is an issue that has been taken up by many studies of system dynamics in the past. For instance, some studies have looked into the effects of workers' attitudes and behaviors (Shin et al. 2014; Gang et al 2012; Jiang et al. 2015) while others have examined the role of managers' commitment to safety (Cooke 2003; Lyneis & Madnick 2009). In Figure 5.2, the safety system is regulated by a balancing loop B1 regarding Management Commitment to Safety and responds to pressures from production and from task scheduling or human error. Poor management practices result in Unsafe Conditions that are inversely proportional to Management Commitment to Safety (MCS). A similar safety model could be developed for workers' commitment to safety and added to the model (e.g., Cooke 2003). However, a more challenging issue would be to model aspects of task scheduling and human error in dealing with taskload due to production demands. For this reason, another model of task management has been developed in a following section to examine the processes of error production and error recovery and their factors that influence their occurrence (Kontogiannis 2011).

The safety model involves a balancing or stabilizing loop B1 where management commitment to safety affects the conditions of work and, hence, the latent failures (Cooke 2003). At the managerial level, the safety loop B1 is affected by Production Pressure that lowers Management Motivation

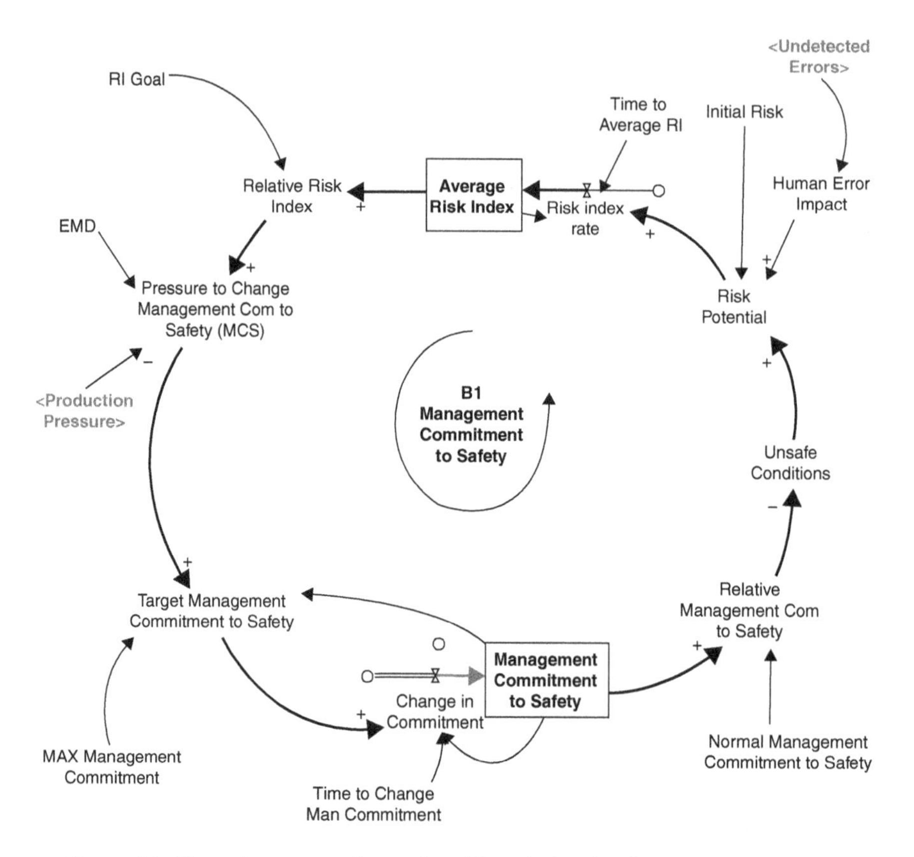

Figure 5.2 The safety system is regulated by a balancing loop B1 representing management commitment to safety

(adapted from Cooke 2003)

to Safety. At the operational level, the safety loop B1 is affected by Schedule Pressure that increases Undetected Errors. These two external variables affect the Risk Index that is represented as a stock variable, averaged over a period of 10 weeks to produce an average value as follows:

$$\text{Average Risk Index} = \text{Initial Risk } (50) + \int_0^t \textit{Risk Index Rate}, \quad (1)$$

$$\textit{Risk Index Rate} = (\textit{Risk Potential} - \textit{Average Risk Index}) / \textit{Time to Average Risk} \quad (2)$$

The Pressure to Change MCS depends on the Relative Risk and the Effect of Production Pressure. Specifically, as production demands increase, the Production Rate Goal exceeds the Order Rate in order to achieve the required inventory, and this leads to a reduction of the Pressure to Change Management Commitment to Safety.

The interplay between commitment to safety and production pressure is formulated in the following equations:

$$\text{Pressure to Change MCS} = \text{Relative Risk} / \text{Production Pressure} \tag{3}$$

$$\text{Relative Risk} = \text{Average Risk Index} / \text{RI Goal}; \qquad (\text{RI Goal} = 10) \tag{4}$$

$$\text{Production Pressure} = \text{Production Rate Goal} / \text{Order Rate} \tag{5}$$

$$\text{Production Rate Goal} = \text{Order Rate} + \text{Inventory Correction} \tag{6}$$

The safety model assumes that management are conscious of any increases of the risk level, which exerts some pressure to increase their commitment to safety. However, their new safety targets are limited by a maximum value that depends on the specific context of work.

Although the safety system has been based on earlier studies (Cooke 2003), this chapter offers a more elaborate algorithm for calculating the Risk Index. The proposed safety model incorporates a human reliability module that takes into account the ways in which tasks are managed by workers and the errors that are not possible to correct in time.

Therefore, a Human Error Impact was calculated as follows:

$$\text{Human Error Impact} = a + b \cdot \text{Undetected errors} \tag{7}$$

The overall Risk Potential can be calculated by taking the product of Human Error Impact and Unsafe Conditions. Given that risk has always an initial value, the current value of the Risk Potential can be calculated as follows:

$$\frac{\text{Risk Potential}}{\text{Initial Risk}} = \text{coef} \cdot \text{Unsafe Conditions} \cdot \text{Human Error Impact} \tag{8}$$

where,

Risk Potential varies in a scale from 1 to 100

Unsafe Conditions vary in a scale from 0 to 1.

Human Error Impact is dependent upon the number of undetected errors that are scaled with an Error Function to create a scale of 1–10

A coefficient is used to make adjustments of the model to real incident data that may exist. In the present study, this coefficient was set at 1

The safety model has been based on the Reason's model (1998), where incidents arise as a result of a combination of latent failures (i.e., management failures) and active failures on the part of sharp-end workers. However, the system dynamic model allows analysts to consider the system interactions that create stabilizing and vicious loops as well as examine the evolution of risks over time.

Production system

One of the advantages of system dynamics is that it is possible to rely on modules created in other scientific domains, such as inventory control and production. In this respect, Figure 5.3 has been based on earlier work on

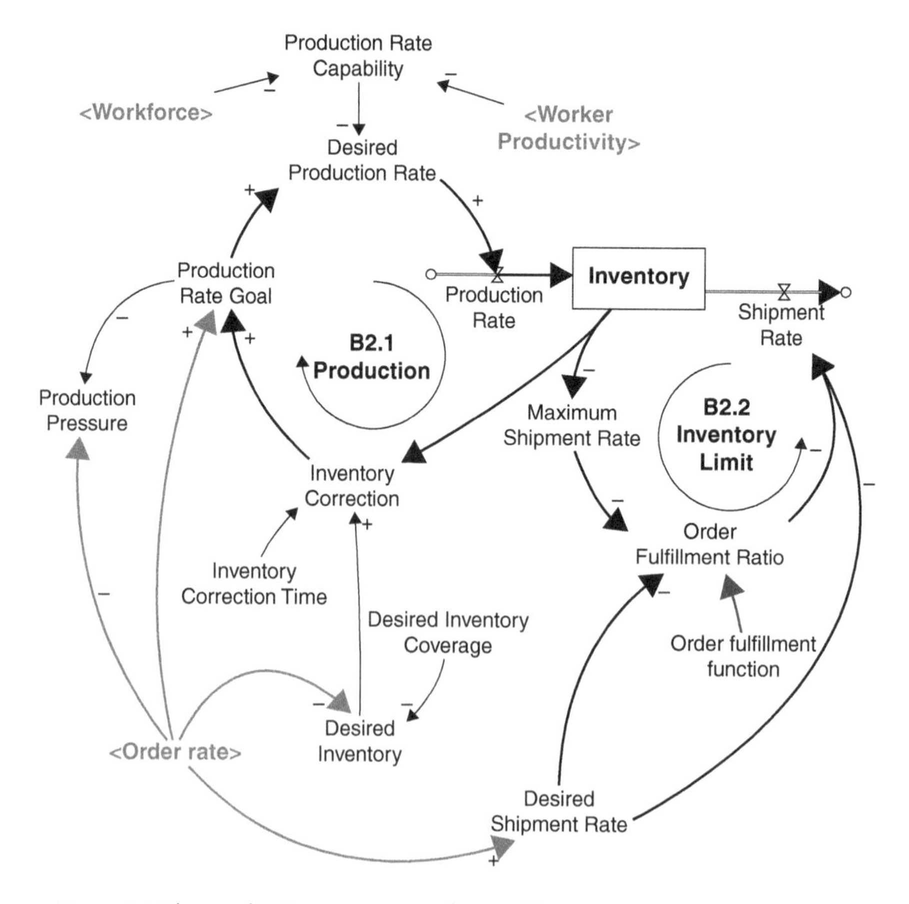

Figure 5.3 The production system as influenced by external variables (i.e., order rate, workforce and worker productivity)

production modeling for the manufacturing and mining industries (Sterman 2000; Morecroft 2007) where inventory level is controlled by the Production Rate Goal that is set to meet the Order Rate of customers and by the Shipment Rate.

The input to inventory level is controlled by the B2.1 production loop (Figure 5.3) which adjusts the Production Rate Goal to meet customer orders. However, this goal is limited by the Production Rate Capability (i.e., Worker productivity times workforce); as a result, a Desired Production rate is used to select the minimum between production goal and capacity. In addition, an Inventory correction factor is used to take into account the difference between the Desired Inventory and the actual Inventory. To be consistent with other studies (Cooke 2003), a four week interval was selected for the Desired Inventory Coverage variable.

The output of the inventory level is controlled with the B 2.2 Inventory Limit loop (Figure 5.3) that prevents a negative inventory value and provides a realistic response to shipment rate when the desired shipment rate gets closer to the maximum. For this reason, an 'Order fulfillment function' has been specified according to the practice in Sterman (2000).

Human resources system

The human resources system regards the processes of hiring new workers, tracking their work experience and estimating their influences on worker productivity. A simple stock and flow structure has been used in Figure 5.4 that regulates the number of workers. The workforce varies beyond the initial number of workers (i.e., 55 miners) according to the difference between the hiring rate of new workers and the attrition rate of existing workers. For simplicity reasons, it is assumed that the newly hired workers would have adequate experience in coping with taskload. A more elaborate model would have to incorporate the regulation of new training programs for new workers.

The attrition rate usually models an organizational problem that appears when experienced workers leave the company because they cannot tolerate an unsafe working environment that is created when the incident rate or the risk index goes higher than normal. In this sense, the adjustment of the workforce is achieved with a balancing loop B3 that hires new workers when there is a demand to increase production significantly.

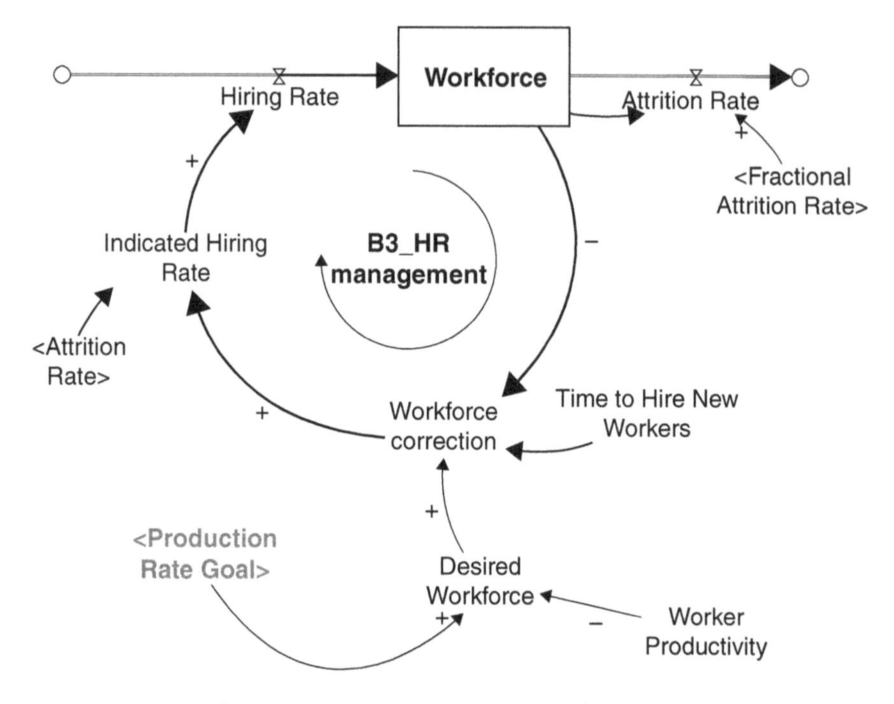

Figure 5.4 The human resources system is regulated by a balancing loop B3

The Hiring Rate is determined by a Workforce correction factor that is calculated by dividing the difference between the desired and actual workforce with the Time to Hire New Workers. The desired workforce can be calculated as the ratio of Production Rate Goal and Worker Productivity. Finally, the Hiring Rate is found by adding the workforce correction factor to the attrition rate of workers. Hence the following equations have been used:

$$Workforce\ Correction = (Desired\ Workforce - Workforce) / Time\ to\ Hire\ New\ Workers \quad (9)$$

$$Desired\ Workforce = Production\ Rate\ Goal / Worker\ Productivity \quad (10)$$

A more elaborate human resources model should treat seriously the issue of worker experience as new workers are unlikely to have similar experience to those who have been many years with the company. Hence, a training program should be designed to raise the experience of new workers to an acceptable standard. This would allow us to address more interactions between management commitment to safety and training.

Task management system

The structure of task management is linked to two balancing loops regarding production requirements and rework of tasks that were not performed satisfactorily. In Figure 5.5, the loop B4.1_Task_Completion controls any increases of Task Backlog, while the loop R4.2_Error_Effects models the side effects of human error. As a result, any increases of Task Backlog may result in higher error rates that further increase task rework and hence, task backlog.

In the B4.1_Task_Completion loop, new tasks flow into the Task Backlog stock, and flow out according to a Task Completion Rate (tasks/week). The Task Arrival Rate (tasks/week) is a function of new tasks required for production and tasks to be reworked due to unrecovered human errors. In turn, the Task Completion Rate depends on the Task Completion Capacity (i.e., potential number of tasks completed per week) that is a function of the number of workers, their weekly hours and the total time per task. However, only a certain factor of the Task Completion Capacity is utilized that reflects the effects of Schedule Pressure on the performance of workers.

The Schedule Pressure refers to the extent that the Task Completion Capacity of the system covers the Total Desired Work Rate (i.e., the work rate of production as required by a certain deadline to complete the backlog). The following equations were used in the B4.1_Task_Completion loop:

$$\textit{Scheduled Pressure} = \textit{Total Desired Work Rate} \,/\, \textit{Task Completion Capacity} \tag{11}$$

$$\textit{Total Desired Work Rate} = \textit{Task Backlog} \,/\, \textit{Deadline to complete Task} \tag{12}$$

$$\textit{Task Completion Capacity} = \textit{Week Shift} * \textit{Workforce} \,/\, \textit{Total time per task} \tag{13}$$

$$\textit{Total Time per Task} = (\textit{Time per Task} * \textit{Experience Effect on Task}) + \textit{Safety Requirements Time} \tag{14}$$

It was assumed that, in order to ensure safe performance, some extra time was required to carry out several preparation tasks and cross checks (i.e., Safety Requirements Time); this extra time would be proportional to the specific production demands.

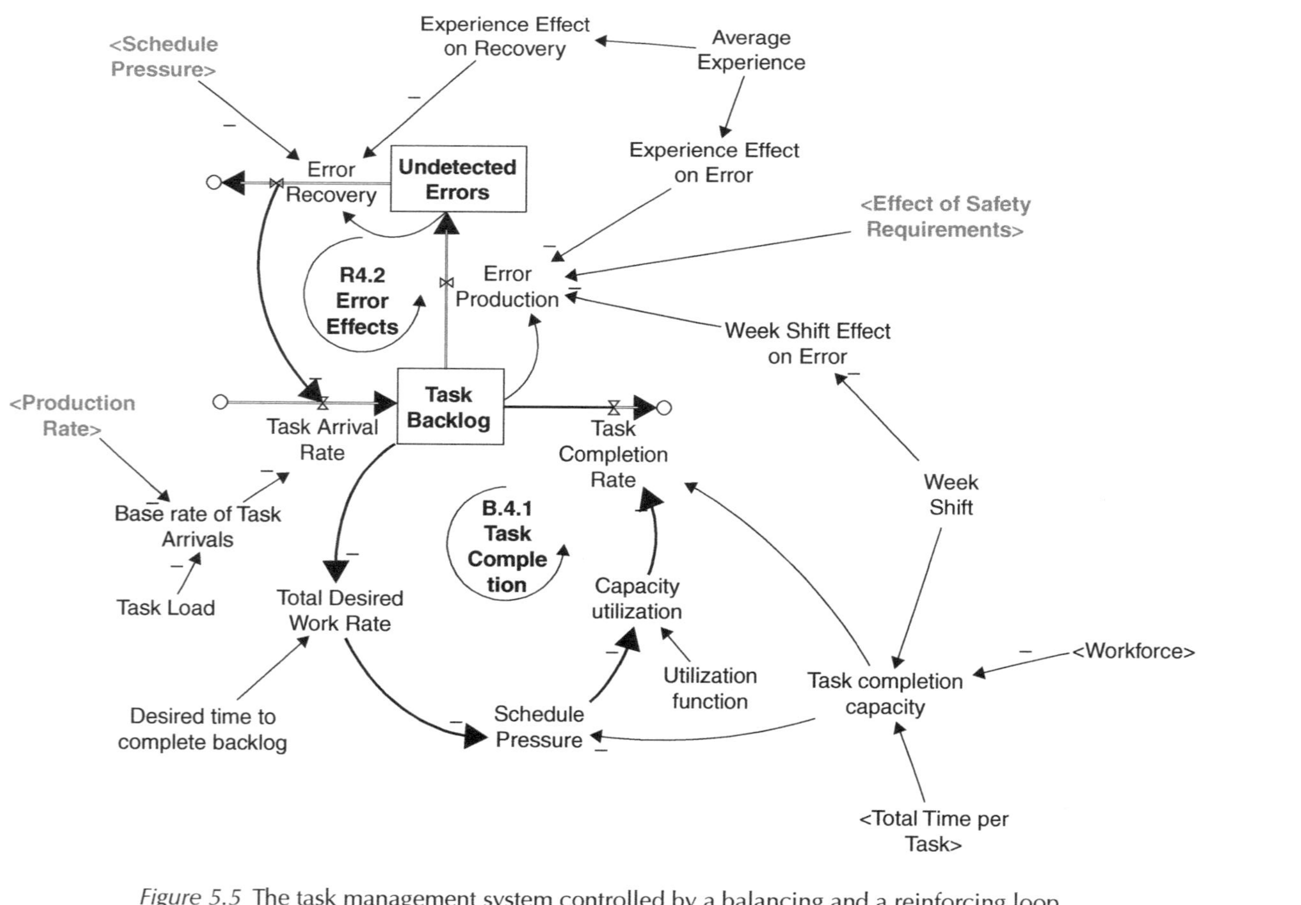

Figure 5.5 The task management system controlled by a balancing and a reinforcing loop

As it can be seen (Eq. 13), the Task Completion Capacity (tasks/week) is affected by the working pattern in the company, which is reflected in the patterns of Week Shift and Worker Productivity. Our model has employed a flexible working pattern where the company requires an increase of work shift by two hours/week/person for a step increase of order rates per 1,000 tons/week; furthermore, it is assumed that worker productivity will increase by five tons/week/person for each step increase in the order rate. As a result, the Task Completion Capacity is maintained close to the Total Desired Work Rate and the Schedule Pressure becomes relatively manageable.

Workers have to manage their own errors, which may result in requirements for reworking tasks which increases the Risk Potential of operations. The number of Undetected Errors is a stock variable that is affected by the processes of error production and error recovery. The number of errors produced is affected by fatigue (i.e., Week Shift Effect on Error), average experience and Safety Requirements per task. Errors may be recovered when experience gets higher and the Schedule Pressure gets lower. Hence, the following equations model the two error processes:

$$\begin{aligned}&\textit{Error Production Rate} = (\textit{Task Backlog} * \textit{Week Shift Effect}) / \\ &(\textit{Experience Effect on Error} * \textit{Effect of Safety Requirements})\end{aligned} \tag{15}$$

$$\begin{aligned}&\textit{Error Recovery} = \textit{Undetected Errors} * \textit{Experience Effect on} \\ &\textit{Recovery} / \textit{Schedule Pressure}\end{aligned} \tag{16}$$

From the perspective of human reliability, it is important to model the processes of error production and error recovery. System dynamics are very useful for modelling the influences of many work factors and for building simple or complex models of human errors. With regard to the error detection process, the model takes into account the number of errors produced (i.e., undetected errors) and the effects of schedule pressure and average experience. As production demands change, so does the schedule pressure which may impact error recovery in a negative way. The effect of experience of error recovery needs to be normalized since experience has a different scale from schedule pressure. For simplicity reasons, the process of error production takes into account the amount of work to be done (i.e., task backload) and the effects of experience, fatigue and safety requirements. As production demands increase, it is assumed that employees work overtime which may give rise to a gradual increase of weekly shifts (i.e., from 40 to 50 hours per week).

Results

Although production demands may take several forms, in our simplified model, variations in demands were seen as step increases in customer order rates while overt responses of worker and management action were entirely effective in bringing the risk index back to an acceptable level. The model has been tested in five increases of customer order rates from 5,000 to 10,000 (in steps of 1,000 tons/week) and it was allowed to run up to a period of up to five years to observe its long run behavior. The model was first tested in a stable situation where customer orders were constant at 5,000 tons/week and where management set an ambitious target to decrease the risk index from 50 down to 10. The results of the model run (Figure 5.6) showed an exponentially decay function of the risk index that reached a stable value of 15, after a year. Although a bit higher than the target risk index, the final value is acceptable since an assumption was made that the target could be achieved without increasing the workforce or the productivity of miners; hence a small number of errors and rework tasks continued to exist.

As orders followed a step increase of 1,000 tons/week (at week 100), the system maintained the risk index at the same value as Management Commitment to Safety increased from 100 to 120. However, the system could not manage a step increase to 2,000 tons/week without a further increase of risk since commitment to safety was restricted by a maximum value of 130. Although, the management responded by increasing the workforce from 55 to 70 persons, and miners increased their productivity from 90 to 100 tons/week (Figure 5.7; production at 7,000 tons/week) the overall risk index almost returned to the original value (i.e., just below 50).

To test the model at extreme conditions, a request for increasing production to 10,000 tons/week was simulated at week 100, resulting in a high risk index of 90 (Figure 5.6). This increase in the risk index was due to an increase of the number of undetected errors (i.e., a higher Schedule Pressure) and an increase in the unsafe conditions (i.e., a higher Production Pressure). This adverse effect gets even higher because of the operation of two reinforcing loops (Figure 5.1) where an increase in the Production Rate Goal is reinforced through the process of worker attrition, which reduces the inventory level and requires a further increase of the Production Rate Goal.

A sensitivity analysis of the effect of Management Commitment to Safety showed that the risk index could drop from 90 to 75 by increasing management commitment to 160. However, this is not always easy to do since

Production order rate at 5000 tons/week at week 100

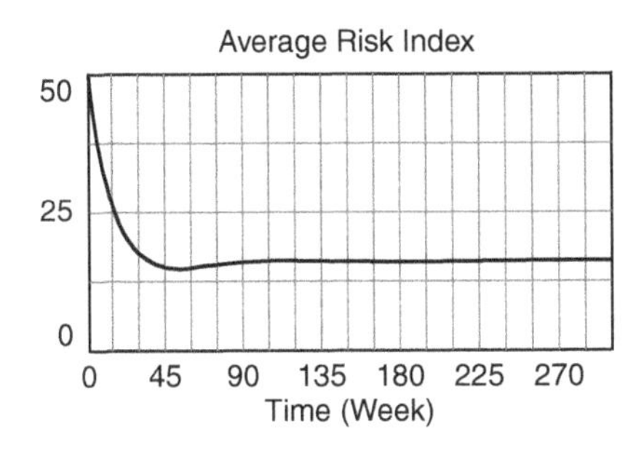

Production order rate at 7000 tons/week at week 100

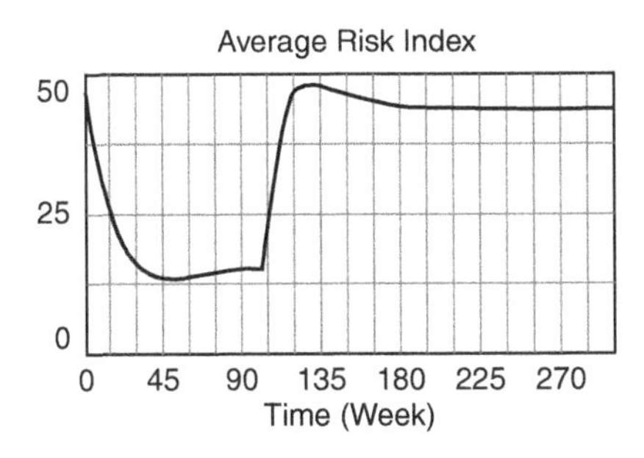

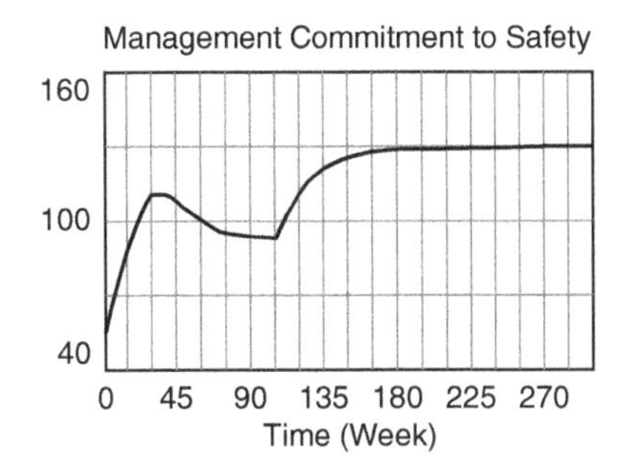

Production order rate at 10000 tons/week at week 100

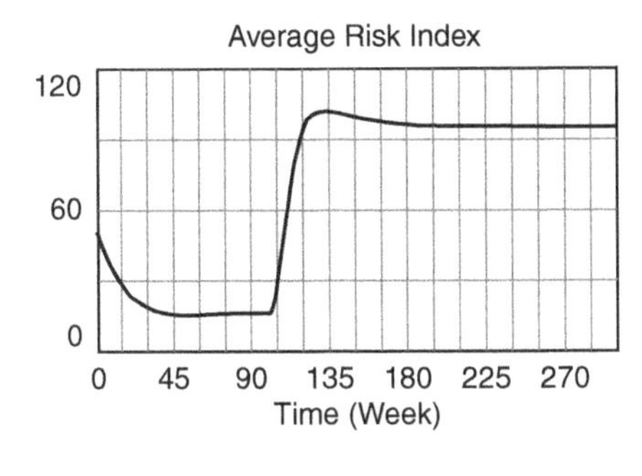

Figure 5.6 The effect of production increases at week 100 on commitment to safety and Average Risk Index (maximum commitment to safety was set at 130)

management is pre-occupied with many production issues, neglecting to raise their commitment to safety. Another policy for managing higher production demands would be to hire a greater number of workers in order to reduce fatigue as well as provide additional training to the newcomers

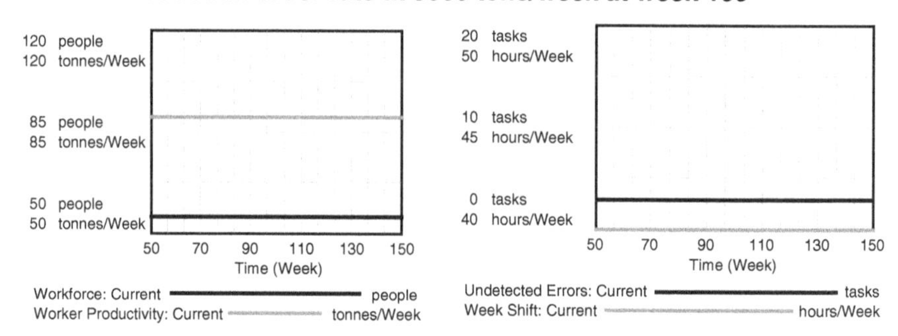

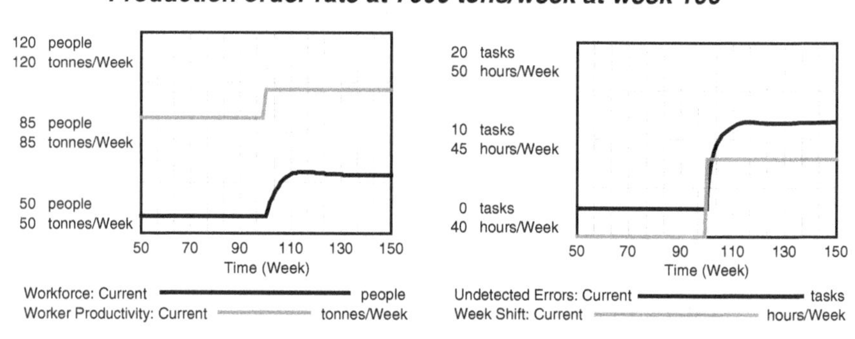

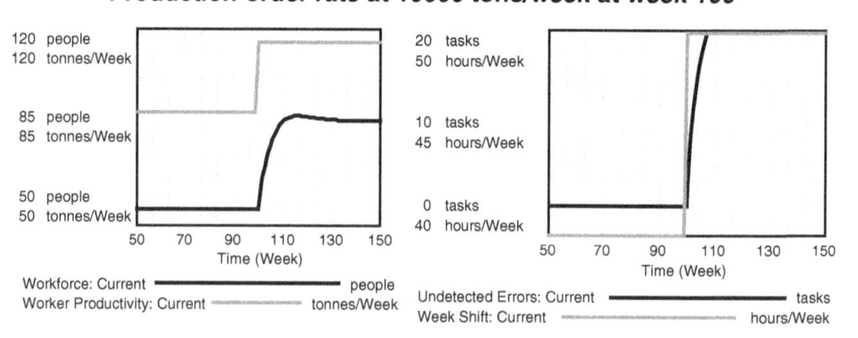

Figure 5.7 Adjustments of workforce, week shift and worker productivity to production increases at week 100 and effects on the number of undetected errors

to ensure adequate human performance. Overall, system dynamics allows analysts to experiment with several system interventions and identify solutions that have the greatest leverage on safety as well as minimize side effects elsewhere in the organization.

Conclusion

The modeling of a typical mine in the case study sought to explore the feasibility of the proposed method rather than perform a realistic numerical estimation of all risks involved in mining operations. Several organizational processes have already been explored by previous studies (Goh et al. 2012; Kontogiannis 2012) and, hence, they were not further discussed in this chapter. The results of the simulation model are open to criticism for the validity of assumptions regarding human and organizational performance since many of the mathematical relationships have not been tested empirically. In the present model, the causal relationships have been grounded in earlier system dynamics studies and most notably the work of Cooke (2003) which was based on the public inquiry of the Westray Mine incident. Furthermore, many of the human and organization patterns of behavior have been based on earlier work on organizational archetypes as applied to system safety (Marais et al. 2006; Guo et al. 2015).

The present model views safety within the overall framework of managerial performance in managing a variety of subsystems, each with its own priorities, demands and constraints. This is in contrast to more traditional approaches where incidents were seen as the result of failures of individual elements such as workers, supervisors, procedures or equipment. Apart from management commitment, there are also other means of enforcing a 'putting safety first' policy that include: providing adequate training to new miners, encouraging workforce involvement that would increase their productivity, maintaining work shifts below fatigue limits, and so on. System dynamics offer a proper methodological framework for understanding how big and how soon analysts can observe the results of safety-related decisions. Organizations can build their own models of safety management and use them as tools for further improvement. As more data are produced for causal relationships in the organizational structure, the easier it will be to attempt empirical validations of system dynamics models.

References

Bouloiz, H, Garbolino, E, Tkiouat, M, Guarnieri, F 2013, 'A system dynamics model for behavioral analysis of safety conditions in a chemical storage unit', *Safety Science*, vol. 58, pp. 32–40.

Cooke, DL 2003, 'A system dynamics analysis of the Westray mine disaster', *System Dynamics Review*, vol. 19, pp. 139–66.

Gang, H, Guo-tong, Q & Tian-bo, J 2012, 'Systematic analysis of impact factors and level of coal miners' safety behavior', *Journal of Business Management and Economics*, vol. 3, pp. 142–6.

Goh, YM, Love, PE, Brown, H & Spickett, J 2012, 'Organizational accidents: A systemic model of production versus protection', *Journal of Management Studies*, vol. 49, pp. 52–76.

Guo, BW, Yiu, T & González, VA 2015, 'Identifying behavior patterns of construction safety using system archetypes', *Accident Analysis & Prevention*, vol. 80, pp. 125–41.

Jiang, Z, Fang, D & Zhang, M 2015, 'Understanding the causation of construction workers' unsafe behaviors based on system dynamics modeling', *Journal of Construction Engineering & Management*, vol. 14, pp. 1–14.

Kontogiannis, T 2011, 'A systems perspective of managing error recovery and tactical re-planning of operating teams in safety critical domains', *Journal of Safety Research*, vol. 42, pp. 73–85.

Kontogiannis, T 2012, 'Modeling patterns of breakdown (or archetypes) of human and organizational processes in accidents using system dynamics', *Safety Science*, vol. 50, pp. 931–44.

Leveson, NG 2011, *Engineering a safety world*, MIT Press, Cambridge, MA.

Lyneis, J & Madnick, S 2009, 'Preventing accidents and building a culture of safety: Insights from a simulation model', *Working Paper WSD-WP-2009-2*, Engineering Systems Division, Massachusetts Institute of Technology. Cambridge, MA.

Marais, K, Saleh, JH & Leveson, NG 2006, 'Archetypes for organizational safety', *Safety Science*, vol. 44, pp. 565–82.

Morecroft, J 2007, *Strategic modeling of business dynamics*, Wiley & Sons, New York.

Reason, J 1998, *Managing the risks of organizational accidents*. Routledge, London.

Shin, M, Lee, HS, Park, M, Moon, M & Han, S 2014, 'A system dynamics approach for modeling construction workers' safety attitudes and behaviors', *Accident Analysis & Prevention*, vol. 68, pp. 95–105.

Sterman, JD 2000, *Business dynamics: Systems thinking and modeling of a complex world*, McGraw Hill, Chicago, IL.

PART

3

Establishing and monitoring a Common Operational Picture

Developing a risk register to deliver risk intelligence

Maria Chiara Leva and
Richard Sheehan

Introduction

This chapter illustrates the necessary steps to develop a risk register and the benefits it can deliver. The concept is also presented though the use of a case study developed in an energy generation company. A risk database, or risk register, is a tool to monitor, review and reduce risks in a n organisation. It's not only a repository but also a knowledge management tool for all the hazards connected to operations and assets in an organisation. It is typically used to manage all the information describing the risk. It has been deployed in several types of industries with the purpose to support safe processes or to support safe and efficient project management. The key role of the risk registers is documenting the process for reducing risk and introducing mitigations in order to maintain safe operation. The example showed in this chapter is around the key steps aimed at developing and implementing a risk register data structure supporting consistent hazard identification and moving into the space of option evaluation. All those aspects should place risk registers at the heart of Total Safety Management.

Risk register: scope and definitions

To maintain safe operations organisations are required to continuously review and monitor their risks. This means that the results of safety studies and/or the evidence of issues collected from operational experience must be translated into a format that can be analysed, reviewed and acted upon, and new data about the level of risk continuously collected to keep the safety information up to date (Monferini et al. 2013). This helps to create an "informed culture", defined by Reason (1997) as a culture in which both management and operators are informed of and knowledgeable about the factors that influence safety as a whole. When the available information is shared between all applicable levels of the organisation, a Common Operational Picture (COP) can be created as the basis for safe and reliable system operation (Kontogiannis, Leva & Balfe 2017; Leva et al. 2014). The development and implementation of a risk register is a way to manage the necessary knowledge and create a common operational and actionable picture for the risks shared across an organisation.

A risk register is a central tool to monitor and reduce risks, both those identified during initial safety assessments and those emerging during operations (Whipple & Pitblado 2010). The risk register should support the prioritisation of the areas that require managerial attention. It typically contains information describing each risk, an assessment of the likelihood and consequences, a ranking according to a risk matrix, the risk owner and information on the mitigations to be put in place. These data can be analysed to present the risk profile for different aspects of the organisation (Filippin & Dreher 2004). When reviewed and updated over time, it can also present trends within the risk profile and focus management attention on activities and facilities presenting higher and or increasingly high risk profiles(Whipple & Pitblado 2010).

Risk registers are used in a variety of industries, e.g. medicine (Brown 2004) and construction (Dunović, Radujković & Vukomanović 2013), as well as high hazard industries such as oil and gas (Hasle, Kjellen & Haugerud 2009) and electricity generation (Leonard 1995). Cooke-Davies (2002) found that sustaining a visible and fit for purpose risk register is one of the key success factors for project management. Patterson and Neailey (2002) highlight the importance of the risk register and suggest that the benefit of a risk register is as a method to enable all stakeholders to "consciously evaluate and manage the risks as part of a decision making process" (p. 365). They also note the importance of the risk register in documenting the process of reducing risk and introducing mitigations. However, Kutsch and Hall (2010) warn of the danger of irrelevance; which is to say the risk registers can easily become a "tick-box" exercise when the owners and contributors do not have a real ability to influence the risks. Despite the clear importance of risk registers in the risk management process, there is very little guidance on their development and implementation (Dunović, Radujković & Vukomanović 2013). This chapter attempts to address this gap in guidance by describing the results of a case study in which a risk register was established in an electricity generation company across multiple locations (see Leva et al. 2017; Balfe et al. 2014). The single central risk register is aimed at collating risks from across the business, including various power stations across different geographical locations and supporting management decisions on corrective actions. The objectives that we will be covering are:

- To develop a risk register data structure supporting consistent hazard identification and risk rating across different sites
- To develop equivalent severity and frequency scales for different loss types and for application across different business units, such as operations, maintenance, finance, etc.
- To use the risk register to highlight key business risks to senior management
- To use the risk register to gather information about mitigation measures in place and their effectiveness
- To embed the risk register within a risk management process and share good practices across the company

Developing a risk register

Key components

Although Risk Registers may vary a lot in shape and content there are some key components that are considered necessary. First is the description of the risk, and a unique identification number to facilitate tracking. A concise description is necessary to allow users and reviewers to understand what is being documented. A more comprehensive description may also be provided, particularly for complex risks or those that have a long history. Each risk must have an indication of its priority, in the form of a risk ranking. Risk rankings are typically calculated from the product of the severity and likelihood of the risk. The calculation may be more or less sophisticated, depending on the data available. Finally, the actions required to improve or manage a risk should be documented, along with the overall risk owner who is responsible for ensuring progress of the risk against the planned timescales (dates). The risk owner may not be responsible for the individual actions but they are responsible for ensuring overall progress. Complex or detailed actions may be held in a separate document, but a summary should always be available in the risk register. The core components of a risk register are reported in Table 6.1.

Table 6.1 Key components required in a risk register

Element	Description
Risk ID	A unique identification number for each risk
Risk Description	A concise description or title for the risk
Risk ranking	A quantification of the risk, based on severity and likelihood
Owner	The person responsible for managing the risk and ensuring actions against it are completed
Actions	A list of actions for each risk
Dates	The date of entry and modification should be held for each risk to assist with reviews. Action target and completion dates should also be included

Other possible elements to be found in a risk register may include documentation of existing controls in order to assist with monitoring their enforcement and effectiveness, the risk status (e.g. open, closed, increasing, decreasing, etc.) to assist with tracking the overall risk profile, the type of risk and associated losses (e.g. safety, financial, reputational, legal, etc.) and a classification for consequences/loss categories, root causes and hazards.

Consequences/loss categories

The division of consequences in categories is one of the central functions of the risk register for helping to judge where money should be mainly invested. The company for which the case study presented in this chapter was developed considered five principal areas where the monetised risk exposure had a relevant impact on the general funds of the company. Therefore, the categories of losses covered by the risk register are not only the ones explicitly linked to financial implication but also the one covering safety, environment, technical performance and reputational effects.

The likelihood categories

It is necessary to estimate how often an event on annual base occur to quantify the risk. The likelihood scale used by the partner company includes probability ranges to estimate corresponding classes of monetised risk values to be able to link risks that appear across multiple stations with different likelihood and severity in the various consequences categories. Both the likelihood scale and the loss categories are based on the MIL-ST-882. This

Table 6.2 Likelihood scale used in the case study

Rating	Name	Description	Likelihood	Mean value of range
1	Unlikely	So unlikely, it can be assumed occurrence may not be experienced	<0.01%	Mean value 0.005%
2	Remote	Very unlikely but possible to occur	0.1%–0.01 %	Mean value: 0.05%
3	Possible	Possible to occur sometimes in the work life	1%–0.1%	Mean value: 0.5%
4	Probable	Will occur several times in the work life	10%–1%	Mean value: 5.5%
5	Frequent	Likely to occur	100%–10%	Mean value: 55%

Military Standard has been recognised as a guiding light in system safety within not only the defence sector but also in transport, energy and aviation (see Leva et al. 2012).

The risk matrix

The risk matrix adopted for the Risk register is there to provide a rough discrete (ordered categorical) approximation to a more detailed, but not readily available underlying quantitative relation between likelihood and severity of scenarios (Risk = probability × consequence). Cox (2008) suggests that such an intuitive interpretation of the risk matrix as an approximation to an underlying quantitative model can only be sustained if the risk matrix, at a minimum, "discriminate[s] reliably between very high and very low risks, so that it can be used as an effective screening tool to focus risk management attention and resources". The use of the matrix is mainly to rank individual risks to allow a better appreciation of their (relative) importance and seriousness. However, as pointed out by Ale, Burnap and Slater (2015) any discussion on the individual acceptability of each risk needs to be done on a case by case basis and generalisations shouldn't be allowed unless the risk estimates associated to the scenario under analysis are supported by further quantification method. The company for which the case study was developed used a 5 × 5 risk matrix with four colours associated with the meanings reported in Table 6.3. However other possible categories of exposure can also be used.

Table 6.3 Categories of exposures used in the risk matrix adopted

Colour	Range of values	Risk level	Comment
Red	15–25	Very high	Unacceptable. Detailed action plan urgently required
Amber	10–14	High	Apply immediate controls
Yellow	6–9	Medium	Apply judgement
Green	1–5	Low	Monitor and manage routine procedures

The hazard categories

The categorisation of hazards depends not only on technical or processing risks but also on all the hazards that involve a losing of money. Then, in a complete and useful risk register it is necessarily categorised as best as possible the hazard in order to have an overhaul framework of the potential hazard like a potential hazard that involves monetised exposure. The hazard categories are:

- Technical: specific hazards relating to equipment
- Process: hazard relating to process
- Work activities: risky activities
- Work environment: hazards relating to the physical plant
- External: all external hazards, including adverse weather conditions, natural hazards or external accidents
- Behavioural: hazards resulting from inappropriate behaviours or use of tools
- Organisational: hazards related to poor support from the organisation
- Environmental: hazards related to the environment, like noise or emissions
- Financial: hazards related to finance
- Project management: hazards related to projects

Aggregation of risks from station level to central level

The rating scheme of the risks based on the risk matrix is used for the purpose of sorting and screening, while the risk register needs to include a further criteria to estimate corresponding classes of monetised risk values to be able to aggregate risks that are in common across multiple stations with different likelihood and exposure in the various impact categories. The risk matrix score is in fact provided on the basis of qualitative scales. Qualitative scales are themselves inherently flawed when it comes to aggregate risk. When using qualitative scales, it is very difficult to say how to compare two High risks with three Medium risks, or how high is a High risk. This is a primary motivation for trying to monetise rating scales.

This is necessary to identify the top ten hazard scenarios relevant across the entire organisation as a whole to facilitate better monitoring by senior management. This will enable hazard categories to be sorted across different stations on the basis of sum of equivalence of economic value of impact multiplied by likelihood for each event category.

The system can ultimately rank risks across 4 economic ranges: they vary from Cat 1 (the highest monetized exposure), Cat 2 (second highest monetized exposure category), Cat 3 (lower monetized exposure category) and Cat 4 (lowest monetized exposure category). A trial implementation of this approach was run in the reported case study. Table 6.4 reports an example of the aggregated top 10 issues obtained using the monetised categories.

Table 6.4 Example of top 10 scenarios aggregated around hazard types at central level with monetised risk ranges (Cat 1, Cat 2, Cat 3 and Cat 4)

ref	top 10 scenarios	Number of affected centres	Monetised risk class	Max CURRENT EXPOSURE	Max FUTURE EXPOSURE
1	Org. Conditions – Inexperienced workers	9	Cat 1	**20**	**12**
2	Fin. – Plant performance	12	Cat 2	**16**	**12**
3	Tech. – Mechanical integrity & design flaws	12	Cat 2	**15**	**12**
4	Tech. – Ageing	11	Cat 2	**15**	**9**
5	Tech. – Reliability	12	Cat 2	**12**	**12**
6	External Safety – Adverse weather	8	Cat 2	**16**	**10**
7	Fin. – Market	12	Cat 2	**16**	**12**
8	Regulatory – Breach of operating license	8	Cat 3	**16**	**9**
9	Regulatory – Civil or criminal litigation	1	Cat 3	**16**	**1**
10	Fire	9	Cat 3	**12**	**10**

Case study: the need and deployment of a risk register in an energy generation company

The case of study is the initial deployment and validation of a risk register for an energy generation company in Ireland and a consistent rating of risks across its stations. The key points for the validations are, first of all, how well the data structure supports the identification and categorisation of risks and how the rating system is used to manage and prioritise risks. Then, another core point of the risk register is surely the attitude of the stations towards use of the final tool and its perceived effectiveness and efficiency when compared with the initial high-level requirements expressed by the organisation. Those requirements are summarised in Table 6.5 below, while Figure 6.1 represents a use case for its deployment.

The feedback was collected in two ways:

1. Through a survey answered by a sample of the main asset specialists in each station in charge of reporting towards the generation risk register
2. Through the feedback collected in the annual risk review workshop held in each station during first quarter of 2014 and 2015

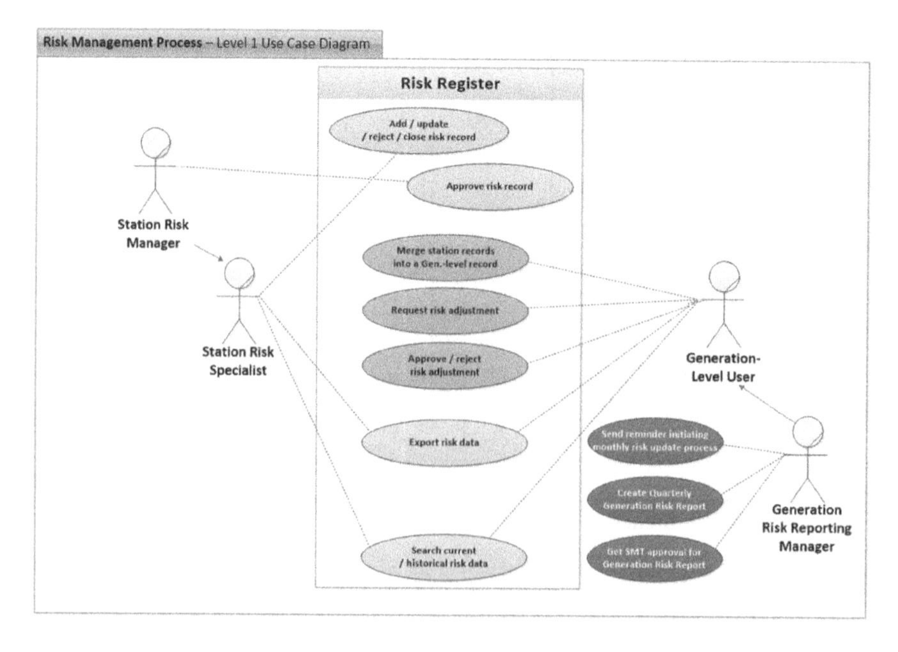

Figure 6.1 Use case for the deployment of the Risk Register in the company

Table 6.5 High-Level Requirements (HLR) for a Risk Register in the company

ID	Description	How it was addressed
HLR1	Create a comprehensive and consistent **risk management process**.	Three new processes were devised – for updating, validation and reporting of risks. Three tiers of stakeholders were defined. Roles of responsibilities were defined. Best practices embedded.
HLR2	Reports/matrices shall provide an update of the **risk levels within the business at a particular moment** of time and take into account possible short term emerging risks i.e. weather extremes, equipment type faults, internal or external incident investigations, etc.	The new central Risk Register caters for different types of risks, based on multiple inputs.
HLR3	The risk management system shall ensure that **all potential hazards** are **identified** and **assessed**.	The Risk Register captures and validates information on potential **hazards** and associated risk **likelihood**, **impact** and **exposure.**
HLR4	The risk management system shall ensure that adequate **control measures** are put in place.	The Risk Register captures information on Existing Mitigations and Additional Mitigations.
HLR5	The risk management system shall ensure that **control measures** remain effective in the management of each risk.	Previously entered mitigations are reviewed and scrutinised as part of the processes around the Risk Register.
HLR6	All risk information shall be held in a **single risk register**, which encompasses all business risks into a single dynamic source.	The Risk Register is a central repository for all risks across the entire business; however, it is manually updated and dynamism is limited. This will improve when full integration with Critical Asset Register is achieved.
HLR7	Periodic hazards and risk reviews shall be carried out.	New Risk Management Processes have been mapped: Monthly Risk Update at station level, Quarterly Validation at Generation level and Quarterly Reporting at management level. In particular, all risks are reviewed at least once a month, and their mitigation plans modified at least: once a quarter (for high and medium scored risks).

(*Continued*)

Table 6.5 (Continued)

ID	Description	How it was addressed
HLR8	Key performance indicators shall be developed to ensure that the key hazards have been identified and assessed, that all business risks are regularly reviewed, and that control measures are in place and effective in reducing risks to a tolerable level.	Tolerable risk levels were defined according to a new standard Risk Matrix as the "Green" (Low) score 1–5 in the 1–25 compound risk exposure scale.
HLR9	The system shall be fully aligned to the company's strategy, with strong and positive management leadership thus ensuring that the business risks are understood from the boardroom to the control room, with real time risk management decision making and a comprehensive risk assessment process for operations.	Detailed process and system guidelines were prepared and distributed to stakeholders and system users to ensure clear understanding and appropriate buy-in.

A key finding from the implementation of the risk register is that the risk register should contain all risks being actively managed – those which require additional investment or further analysis and those which have a high degree of uncertainty associated with them. Low level risks or risks managed on an on-going basis through established business process will not benefit from the additional scrutiny of being on the risk register, and may serve to obscure more critical risks. However, the set of hazards should be as comprehensive as possible, in order to prompt entries and help with analysis of the data held in the register. Some flexibility in the hazard categories may be necessary to allow them to be adjusted and expanded to accommodate changes.

A strong rating scheme for likelihood and severity of each documented risk is the best method currently available for ensuring consistency within the risk register. The scales should be applicable across the business, sensitive enough to collect useful information on the smaller business units but relevant enough to allow comparison across business units and prioritisation of the business risks. However, the nature of risk assessment is speculative and sufficient empirical data is rarely available to accurately quantify either the likelihood or severity of foreseen risks. Such data might be collected over time through data mining accident and incident databases, as well as the risk register itself, but this is currently beyond the ability of most organisations. The rating is therefore somewhat subjective, and even using a clearly defined scale, one user may be inclined towards higher ratings than another. As well as the unintentional variability introduced through subjectivity, some business units might also intentionally increase their ratings to theoretically possible but unrealistic values in order to highlight an issue in their area and attract investment to address it. It is therefore necessary to include a review step in the business process underlying the risk register. This review process may have several aims, but one should be the consistent rating of risks across stations. As issues emerge, additional guidance and modifications can be added to the scales to improve the reliability of the ratings, and any changes should be fed back to users regularly to help improve consistency.

Station attitudes

Initial feedback from the stations has been predominantly positive and uptake has been strong; all stations have contributed risks to the register and regularly update their risks. Feedback on the coding scheme was collected during the station workshops and used to iteratively improve the risk register during the early implementation period. More formal collection of feedback from stations was achieved via a survey, with seven responses (representing almost 60% of the station managers). The survey determined that use of the Risk Register had improved from between quarterly and yearly for the old "Top 10" format to between monthly and quarterly for the new Risk Register. Figure 6.1 describes the overall perception of benefits of the Risk Register (1 = strongly disagree; 5 = strongly agree). The majority of anticipated benefits have been achieved in the eyes of the front line users, but there is room for improvement in terms of supporting periodic reporting in terms of KPIs, the consequence rating scale and accounting for short term emerging risks.

Data analysis supported by the risk register: a few examples

The data analysis for a Risk register is aimed at providing an harmonised view of the main hazardous scenarios shared across various stations. All

Risk Register Benefits

Influencing risk awareness
Data trending of risks
Support periodic reporting against KPI
Helps clarify whether control measures...
Support evaluation of identified control...
Consequence rating is applicable
Frequency rating is applicable
Account for all possible consequences
Accounts for short term emerging risks
Helps make our risk mgt more consistent
Helps identify and plan adequate...
Accounts for all potential risks
Helps senior mgt to feed back to stations
Helps report risks to senior mgt
Helps us to better manage risk
Process communicated
Good overview of risk

1 2 3 4 5

Figure 6.2 Responses collected from evaluation questionnaire

the analyses below are referred to the risks actively managed in the case study, showing the distribution of current risk exposure related to various types of hazards. Figure 6.3 shows the hazards across the stations and their monetised risk exposure extracted from the risk register in 2017, while Figure 6.4 reports the one for 2018. Figure 6.5 offers an example of comparison between the distribution of the hazard "Tech-Mechanical Integrity" for key assets between 2017 and 2018.

One of the examples chosen reports the evolution in time for the technical risk of mechanical integrity.

The data analysis can also focus on the consequences Figure 6.6 shows the comparison of the monetised risk exposure for the worst six type of consequences reported in the risk registers between 2017 and 2018.

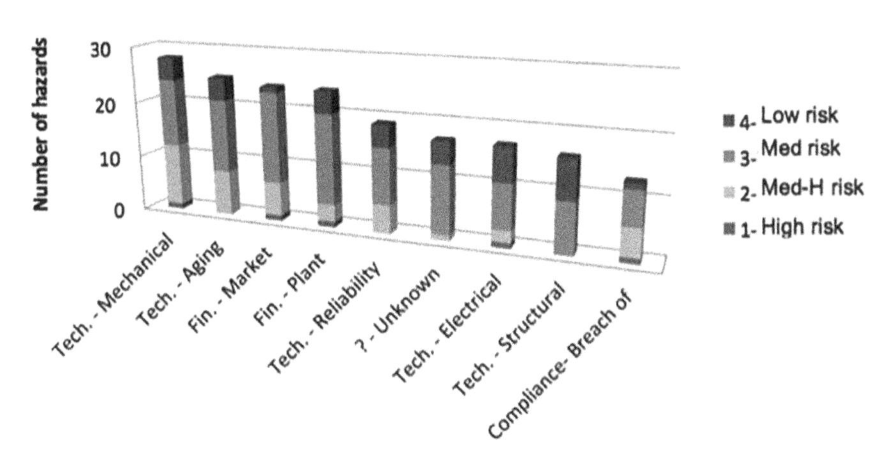

Figure 6.3 Overview of the top hazards reported in the risk registers' 2017 share across multiple stations

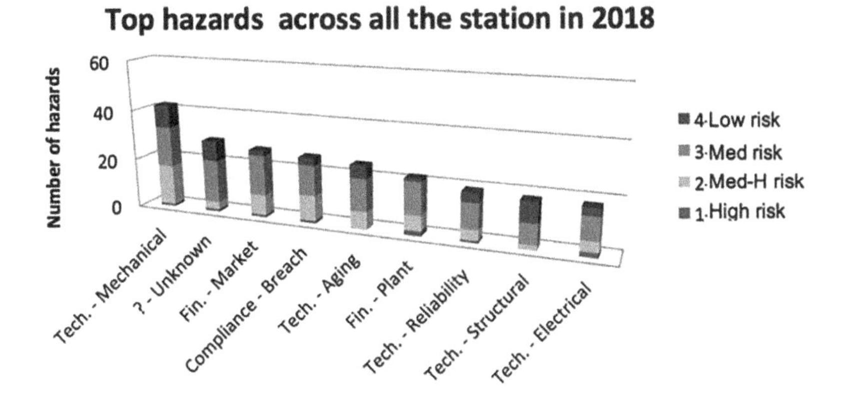

Figure 6.4 Overview of the top hazards reported in the risk registers' 2018 share across multiple stations

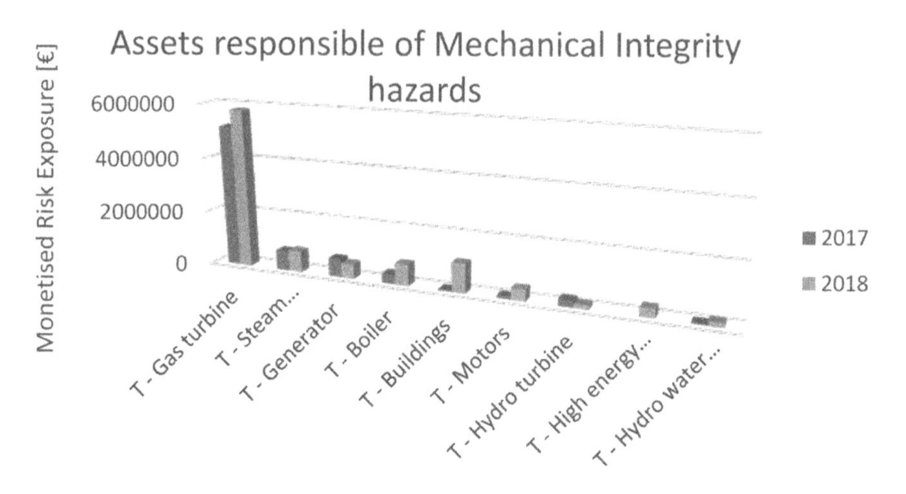

Figure 6.5 Comparison of the distribution of the hazard "Tech-Mechanical Integrity" for key assets between 2017 and 2018

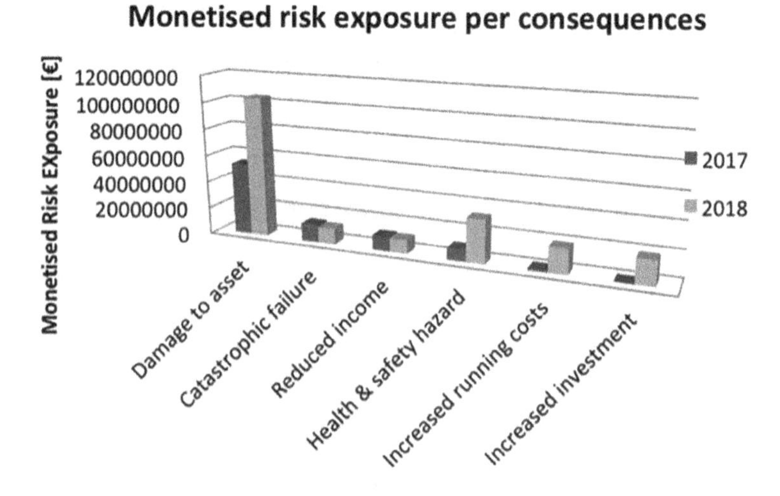

Figure 6.6 Overview of the top consequences and their monetised risk exposure in the risk registers 2017–2018

It is possible to extract, from the risk register, that the responsible hazard for *Damage to asset* in the example above has been *Org. Condition – Inexperienced Workers*. Furthermore, the unique asset involved in this hazard has always been reported as "*H-Organisational conditions*". Finally,

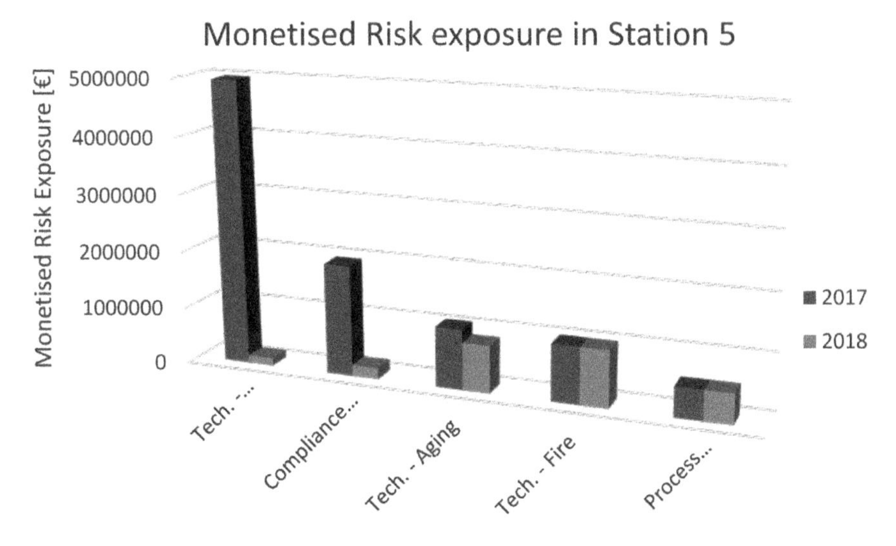

Figure 6.7 Monetised risk exposure trends in Station 5 between 2017 and 2018

the root causes reported for *Damage to asset* consequence are related to human performance deficiencies and worker training deficiencies.

The possibility to perform data analysis on root causes for key hazard/consequence is another useful feature of the risk register used by the company. Each station can also focus on its own trends and progresses (or lack of), see Figure 6.7.

Integration with critical asset registers (CAR)

A component or system in the company is defined as Safety Critical if its function is to prevent an abnormal condition escalating into a major incident (ISO 55001 2014). Within the case study organisation, a major incident is defined as an occurrence (including in particular a major emission, fire or explosion) resulting from uncontrolled developments in the course of the operation of plant, leading to life-changing serious injuries or loss of life, serious danger to the environment (immediate or delayed), extensive damage to property and plant, inside or outside the station. To ensure best practice in terms of safe and efficient asset management the company adopted the industry standard PAS 55 issued by the Institute of Asset Management and published by the British Standards Institution in 2004 (PAS 55 2008). The standard provides guidance across several aspects of good asset management, from lifecycle strategy to everyday maintenance (cost/risk/performance). It was then transitioned to ISO 55000, an international standard covering management of physical assets.

The company currently uses a common structured spreadsheet as a critical asset register (CAR) for all stations. The tool is held in a spreadsheet format stored in an integrated online shared location, with a separate workbook for each station, and is accessible to central asset specialists.

The tool reports the minimum required information for each station and calculates a risk rating for each based on the status, but it does not actually detail the failure modes and their consequences upon which the risk rating is based. The risk register fills this gap. If the CAR tool were to support a better identification of hazards associated to each piece of equipment in alignment with what is required by the risk register, the tool could in itself cover all the requirements to also serve the purpose of a risk register for the company technical assets with customized options for each station to support the management of technical asset risks. This in turn will lead to move also the current CAR towards a Web-based and integrated knowledge management IT solution that can be considered a module of the risk register (or be able to export data directly into it).

The need to support options evaluation with the data of the risk register

The necessity to integrate the Risk Register with the CAR comes from the necessity to manage corrective actions for technical assets that consider both requirements for continuous operation and risk mitigation. This leads to the requisite to support the evaluation of different options of investment for corrective action.

The option evaluation requested from the power generation company has to analyse how an investment can benefit the company in terms of reduction in monetised risk exposures and or benefits achieved throughout the years. This kind of analysis has to be done for the estimated life span of the plant (e.g. 10 years) and then it will be possible to evaluate that the risk rating will increase progressively during the years if investment to reduce the risk will not be done. Then, the option evaluation is aimed at comparing the different option of investment that the company can consider. There are two different kind of investment that the power generation company uses to report:

- Capital Expenditure (CAPEX), which is to say the money a company spends to acquire or upgrade productive assets in order to increase the capacity or efficiency for more than one accounting period
- Operating Expenses (OPEX) such as downtime costs or NDT (Non Destructive Testing)

The evaluation of the best investment option has to compare the monetised risk exposure, adding potential other benefits (e.g. reselling the used asset when the new one has been bought), using a Net Present Value. The NPV is the difference between the present value of cash inflows and cash outflows over a period of time. In other words, it is possible to describe the NPV as the measure of the investment's return. The formula for calculating the NPV is:

$$NPV(i,N) = \sum_{t=0}^{N} \frac{R_t}{(1+i)^t} \qquad (1)$$

where:

- R_t is the cash flow at the time t
- i is the return rate, that is the profit earned from the investment (for the case study of the power generation company a return rate of 8,4% is used)
- t is the time of the cash flow.

Therefore, the comparison between the benefits achievable as a reduction of the total monetised risk exposure plus other possible benefits (derived from continuous operations and or other forms of revenues) and the NPV of the costs to be invested to achieve them can support the management decision about which option should be the best one. The option with the higher Evaluation Index in fact can be considered to be the best compromise between monetised risk exposure and investment. IT can also be used as a function to be maximised if the parameters used from the evaluation can be varied using optimisation algorithms. The evaluation index that can be used is the one reported in formula 2.

$$\text{Evaluation index} = \frac{\sum_{t=0}^{N} (\text{NPV of benefit})}{\sum_{t=0}^{N} \text{NPV of costs}} \qquad (2)$$

Table 6.6 shows an example of five different options of investment for a period of six years with their own evaluation index.

Table 6.6 Example of risk-based option evaluation for corrective interventions

Profile/Option	NPV cost	NPV with benefits	Evaluation Index
Risk Profile (no action)	–	€ −79.336.752	–
Option 1	€ 2.108.222	€ 47.475.450	**22,52**
Option 2	€ 4.711.586	€ 41.885.628	**8,89**
Option 3	€ 11.084.904	€ 45.270.775	**4,08**
Option 4	€ 7.463.701	€ 42.853.385	**5,74**
Option 5	€ 4.531.228	€ 44.058.350	**9,72**

Conclusions

The considerations discussed in the previous section would suggest that the best way to further implement the knowledge management capacity of the risk register is by integrating new functions into the current IT system used in the company for accident incident and near miss reporting by providing a further elements for hazard identification (not retrospective in nature) and to follow up the risk review process and the monitoring (audit) for each station and at the same time facilitating the sharing of best practices and information across the multiple locations (as a Web-based application). The framework reported in Figure 6.8 is a vision of future possibilities for safety and risk management building on the kinds of risk register reported within the case study. This vision describes a situation where existing systems containing relevant information (e.g. asset registers, incident databases, audit system, etc.) are linked to the risk register. Ideally this link would be automatic to reduce manual processing time and/or duplication of effort. Harnessing this information allows a comprehensive and dynamic representation of risk to be developed in the risk register. This information must be reviewed and acted upon at both a local level, to ensure front line risk management, and at a central level to ensure company-wide measures are implemented where necessary. Utilising the inputs and modifications from expert reviews, the central Risk Register tool can identify risk priorities and calculate KPIs for risk management across the organisation, thus enabling effective and efficient monitoring and feedback by senior management.

Figure 6.8 further elements to be integrated in a company Risk Register

The initial research focused on identifying what data already exist, and what information can be generated from that data and has already identified the following elements:

1) Need to integrate with existing IT tool used for accident and incident and process safety events
2) Need to integrate with tool used to collect information and manage workflow from Audits in each station and collect information about effectiveness of Mitigation measures
3) The risk register needs to support better at station level the workflow around the periodic status updates and review of the main company risks and the feedback between stations and central management review of the same risks, especially when central mitigation strategies for aggregated risks may be more effective than local mitigation measures only.
4) Need to integrate with periodic review of operational logs to keep track of how certain risk are evolving in day to day experience
5) The risk register duplicates some of the information the company already collects as part of the critical asset register. If the critical asset register

were to facilitate the collection of the main hazard categories and scenarios associated to asset reporting high risks it would facilitate the establishment of an asset based risk repository that could be generated already as part of the information collected at stations level in the asset register.

Acknowledgements

This publication has emanated from research supported in part by a research grant from Science Foundation Ireland (SFI) under Grant Number 14/IFB/2718 and by the EU FP7 project "Total Safety Management for Critical Activities" (TOSCA; see www.toscaproject.eu) under Grant Agreement FP7-NMP-2012-SMALL-6–310201. The author wishes to acknowledge the assistance of Richard Sheehan, Yilmar Builes, Paride Peritore.

References

Ale, B, Burnap, P & Slater, D 2015, 'On the origin of PCDS-(probability consequence diagrams)', *Safety Science*, vol. 72, pp. 229–39.

Balfe, N, Leva, MC, McAleer, B & Rocke, M 2014, 'Safety risk registers: Challenges and guidance', *Chemical Engineering Transactions*, vol. 36, pp. 571–6.

Brown, AS 2004, 'Finding the hidden risk with medical devices: A risk profile tool', *Quality in Primary Care*, vol. 22, no. 2, pp. 137–40.

BSI PAS 55-2. 2004. Asset Management, Part 2: Guidelines for the Application of PAS 55-1, London: British Standards Institution. ISBN 0-580-42765 X

Cooke-Davies, T 2002, 'The "real" success factors on projects', *International Journal of Project Management*, vol. 20, pp. 185–90.

Cox, LA 2008, 'What's wrong with risk matrices?', *Risk Analysis: An International Journal*, *28*(2), pp. 497–512, Wiley online library.

DoD, U.S. (2012). Mil-std-882e, department of defense standard practice system safety. US Department of Defense.

Dunović, IB, Radujković, M & Vukomanović M. 2013, 'Risk register development and implementation for construction projects', *Gradevinar*, vol. 65, no. 1, pp. 23–35.

Filippin, K & Dreher, L 2004, 'Major hazard risk assessment for existing and new facilities', *Process Safety Progress*, vol. 23, no. 4, pp. 237–43.

Hasle, JR, Kjellen, U & Haugerud, O 2009, 'Decision on oil and gas exploration in an Arctic area: Case study from the Norwegian Barents Sea', *Safety Science*, vol. 47, pp. 832–42.

ISO 55001 2004, Asset Management, BSI Group, viewed 10 February 2014.

Kontogiannis, T, Leva, MC & Balfe, N 2017. 'Total safety management: Principles, processes and methods'. *Safety Science*, vol. 100, pp.128–42.

Kutsch, E & Hall, M 2010, 'Deliberate ignorance in project risk management', *International Journal of Project Management*, vol. 28, pp. 245–55.
Leonard, JB 1995, 'Assessing risk systematically', *Risk Management*, vol. 42, no. 1, pp. 12–17.
Leva, MC, Balfe, N, McAleer, B & Rocke, M 2017. 'Risk Registers: Structuring data collection to develop risk intelligence'. *Safety Science*, vol. 100, pp.143–56.
Leva, MC, Balfe, N, Kontogiannis, T, Plot, E & Demichela, M 2014, 'Total Safety Management: What are the main areas of concern in the integration of best available methods and tools?', *Chemical Engineering Transactions*, vol. 36, pp. 559–64.
Leva, MC, Pirani, R, De Michaela, M & Clancy, P 2012, 'Human factors issues and the risk of high voltage equipment: Are standards sufficient to ensure safety by design?', *Chemical Engineering Transactions*, vol. 26, pp. 273–8.
Monferini, A, Konstandinidou, M, Nivolianitou, Z, Weber, S, Kontogiannis, T, Kafka, P, Leva MC & Demichela, M 2013, 'A compound methodology to assess the impact of human and organizational factors impact on the risk level of hazardous industrial plants', *Reliability Engineering & System Safety*, vol. 119, pp. 280–9.
Patterson, FD & Neailey, K 2002, 'A risk register database system to aid the management of project risk', *International Journal of Project Management*, vol. 20, pp. 365–74.
Reason, J 1997, *Managing the risks of organisational accidents*, Ashgate, Aldershot.
Whipple, T, Pitblado, R 2010, 'Applied risk-based process safety: A consolidated risk register and focus on risk communication', *Process Safety Progress*, vol. 29, no. 1, pp. 39–46.

Continuous monitoring of safety performance

Payam Amir-Heidari

Glossary of abbreviations

DFF	Deviation Frequency Factor
DI_{lag}	Lag Deviation Index
DI_{lead}	Lead Deviation Index
DSF	Deviation Severity Factor
SPI	Safety Performance Indicator
SPI_{lag}	Lag Safety Performance Indicator
SPI_{lead}	Lead Safety Performance Indicator
MAL	Minimum Acceptable Level
MKF	Missing KPI Factor
MSW	Minimum Significant Weight
RF	Reward Factor

Introduction

Determining a mechanism for performance measurement and monitoring is a key part of any management system, and this has a higher importance for total safety management. There is an old belief that says: 'if you cannot measure something, you cannot manage it' (Hendershot 2007). Key performance indicators (KPIs) are the fundamental building blocks in measuring and monitoring processes of safety management.

KPIs have been used to monitor the progression over time within a company or to compare results between companies (Swuste et al. 2016). The information provided via performance indicators can be used to facilitate the implementation of appropriate risk management actions (Haas & Yorio 2016). In the following sections, a classification of safety KPIs and a review of literature leads to a new framework that is described in detail in this chapter.

Safety is a dynamic qualitative condition of a system and is only measurable indirectly by proxies or indicators (Swuste et al. 2016). Generally, safety can be divided into process safety and personal (or occupational) safety.

Process safety deals with the prevention of 'process events' that are associated with hazardous substance releases (e.g. release of toxic, flammable, corrosive and active materials) and/or energy releases (e.g. rupture of compressed gas). Therefore in process safety, prevention of LOCs (loss of containment) and asset integrity are major concerns (Hassan & Khan 2012). Process accidents usually have high severities and low frequencies. Their occurrence is related to design defects, lack of mechanical integrity, defects in safety systems, nonconformity with working procedures, human errors, etc.

Personal safety deals with the prevention of occupational accidents that usually have low severities and high frequencies. When they occur, the adverse impact involves only a few persons. Falling from a height and amputation are two examples of personal safety accidents. From a personal safety viewpoint, unsafe practices and unsafe conditions are the main causes of accidents.

It has been shown that there is not a strong relationship between good occupational safety performance and process safety performance, so indicators of occupational safety do not necessarily have a relationship with process safety. This means that process safety indicators and personal safety indicators should be defined separately.

In a unit or plant, process accidents (e.g. tank rupture, toxic release, etc.) are usually rare, so if there is no record of process accidents in the past few years, one cannot say that the process safety management system is in accordance with the standards expected by society. Therefore, for process safety the use of lead indicators is vital. Process safety leading indicators are classified in three categories (Rogers & Pasman 2013): mechanical integrity indicators, action items follow-ups and training/ competence indicators.

Past experience

The measurement of safety has been rooted in the quality measurement domain that has been an interesting area of research in service quality and health care quality measurement. In addition, performance measurement has been an active area of research in business management. The definition of 'indicators', 'measures', 'proxies' or 'metrics' is a key point of quality and performance measurement. Similarly, in safety performance, recent studies have focused on key performance indicators (KPIs).

In the past, researchers have used different methodologies to identify and select safety performance indicators. Juglaret et al. (2011) suggested that because health and safety management systems are made up of several interacting elements that form a unit cycle, indicators should be selected to provide information on the distinct effectiveness of each element. These elements include change management, leadership development, contractor management, emergency preparedness, etc. A similar approach for identification of safety KPIs has been recognized in other researches (Podgorski 2015; Haas & Yorio 2016).

Wu et al. (2015) provided insights into the cause-effect relationships between leading safety indicators and goals in the construction industry. The authors explained that the level of safety performance in construction sites is mainly affected by four elements: safety climate, safety culture, safety attitude and safety behaviour. Based on literature review, they determined 26 key indicators to measure these four elements. In that work, Structure Equation Modelling (SEM) was used to determine the degree of importance of these elements. This methodology was used for a comparison of safety performance across different sectors of the construction industry. In that methodology, the final aggregated performance index was a vector in fuzzy format. Despite the strong theory, this method is relatively complicated and time-consuming, so it may be inappropriate for continuous use in every kind of company.

Hassan and Khan (2012) presented a hierarchical structure for the monitoring and assessment of asset integrity indicators, which are closely related to process safety. In that work, a so-called risk metric was used to classify asset integrity through the integration of leading and lagging indicators. Expert opinion and the Analytic Hierarchical Process (AHP) were used to determine the weights of indicators, and to aggregate them to determine an asset integrity risk level or a 'risk-based asset integrity indicator'. After creating a hierarchical structure of indicators, a 'risk factor' value was assigned to each specific indicator, beside its weight. The risk factor for leading and lagging indicators was defined separately. For a lagging indicator, it was equal to the product of event frequency and its consequence severity, while for a leading indicator, it was calculated by multiplying the related fail percent (100% – % of success) with the importance of that indicator. Next, an aggregation was performed by applying the weighted arithmetic mean of 'risk factor' values for each level in the hierarchical structure. In that methodology, the mean value of each level becomes the input for its root in the upper level, and through a bottom-up process the final value of 'risk-based asset integrity indicator' is determined. After aggregation, a guide table or criteria table was used to evaluate the calculated indicator level.

In spite of different methods used for measuring safety performance in the past, there are yet some fundamental discussions indicating ambiguities in the classification and the use of different types of KPIs. Most of these discussions were presented in a critical paper published by Hopkins (2007). Because of the importance of the discussions, a brief categorization of safety KPIs is provided in Table 7.1 based on a review of the literature.

The classification based on the 'scope' and the 'time' has been discussed broadly in the previous studies. The 'type' categorization is rather new, and it was introduced by Hopkins (2007), who distinguished between two concepts related to lead KPIs: 'input measure' and 'output measure'. For example, consider the percentage of equipment that is past due for inspection for active monitoring. Some indicators may relate to the monitoring being carried out in a timely fashion (e.g., percentage of tests conducted by due date). Other indicators may relate to the results of the monitoring process (e.g., percentage of cases in which equipment fails the test). The first is a measure of monitoring activity, while the second is a measure of equipment adequacy. The first can be termed an 'input measure' and the second as 'output measure'.

Table 7.1 Categorization of safety KPIs based on different criteria

Criteria	Categories	Example
Scope	Process Safety	Rate of LOCs (loss of containment)
	Personal Safety	Rate of falls
Time	Lead	Percentage of equipment passed mechanical integrity tests successfully
	Lag	Rate of LOCs (loss of containment)
Type	Input	Percentage of technical inspections conducted by due date
	Output	Percentage of inspection cases in which equipment fails the test

In determining the performance of an organization, this implies that the lack of complete measurement of a special KPI or missing data about a KPI should have a negative effect in performance level. However, if the KPI is measured truly by the organization, then the level of the measured KPI should be taken into account in examining safety performance. This idea that was adopted from Hopkins (2007) was utilized in our framework of safety monitoring (by defining *MKF*), as described in the methodology that follows.

The proposed methodology

In designing a framework for measuring safety performance, the first step involves a method for choosing proper indicators that reflect accurately safety performance. Interestingly, indicators seem to be mainly based on experience from companies or on common sense (Swuste et al. 2016). In the past, literature reviews and expert opinions have been used for the identification and selection of KPIs (Podgorski 2015; Antao et al 2016; Haas & Yorio 2016; Wu et al. 2015; Hassan & Khan 2012).

Use of proper KPIs is critical for the effective measurement and monitoring of safety performance. We need to decide on what the indicators are for and set out specific criteria for what makes a proper indicator for specific purposes. By analogy with measuring instruments, features that should be considered in selection of good indicators can include: validity, reliability, sensitivity, representativeness, openness to bias and cost-effectiveness (Hale 2009).

After selecting the criteria for proper indicators, the second stage involves making decisions about the number of indicators and how to deal with them. Some studies have developed and used a large number of indicators (i.e., more than 400 indicators). In practice, the use of a large number of indicators has many disadvantages since senior management usually needs a limited number of indicators (Rogers & Pasman 2013). In order to create a relatively small set of KPIs on the basis of a larger initial set of candidate indicators, two basic approaches are possible: selection and aggregation (Podgorski 2015).

Selection refers to choosing the most significant and representative indicators out of a relatively large number of potential indicators. Usually, multi-criteria decision-making (MCDM) analysis methods have been used for prioritizing and selecting important indicators. Aggregation involves determining the value of a higher-order performance indicator, with the aim to reflect values of all lower-order indicators (sub-indicators) in a collective and synthetic manner. A common method of aggregation is the use of the arithmetic or linear mean. The KPIs received in the course of aggregation of sub-indicators have been previously used for the measurement of safety of nuclear power plants (Podgorski 2015).

A combination of selection and aggregation is probably the best method to be used in the design of a good framework for safety performance measurement. The selection process can be based on expert opinion using existing sound methodologies, such as AHP (Saaty 1987). However the aggregation process may need an exact quantification philosophy and a rational mathematical model to produce a sensitive integrated index from the data of a large range of different KPIs.

One of the greatest problems with indicators is that managers learn to manipulate them, which contributes to a false sense of security (Hale 2009). Therefore, development of a structured strong mathematical framework for performance measurement can help us to solve this problem. The proposed framework in this chapter uses a combination of qualitative and quantitative methods to determine safety performance, so it has a semi-quantitative structure. According to this framework, the safety monitoring process is composed of three stages that are described in the following sections.

Stage I: qualitative study of the safety system

The first step in design of a performance measurement system is creating an expert team with deep knowledge and experience in the field of safety and KPIs. The first stage of the framework is expected to direct the expert team to gain qualitative information about the organizational environment and its safety system. In this stage, qualitative knowledge about the internal and external environment of the organization is gathered, and its activities and processes/operations are recognized. Next, general information about the status of safety system in the organization can be gathered. In this stage, the organizational structure of the safety system is identified and a qualitative assessment of its performance is performed. Inputs to this stage can include recordings of the past accidents and losses. Also, a review of past hazard identification and risk assessment reports can be very helpful. If there is not enough data, it is highly recommended that a baseline comprehensive risk assessment be carried out to find out the major risks and main problems. In this stage, a general picture of the organization and its safety system is gained and its qualitative safety performance is assessed. Using this qualitative knowledge and existing guidelines and documents about safety KPIs, an extended list of KPIs can be developed that is applicable to the whole organization. The KPIs in this list must be categorized based on time factor, as described in the section "Past experience".

Stage II: KPIs selection, rating and evaluation criteria

In stage II, the leading and lagging KPIs are separated and arranged in two groups. The important KPIs should be selected by an expert team. If the decision of team members about important KPIs of each class does not converge, qualitative decision making methods like Delphi (Dalkey 1969) can be employed. Next the KPIs of each group are weighted and prioritized separately with the AHP method. The team should decide on the number of KPIs in each group and select the final sets of leading and lagging KPIs (a maximum number of 15 KPIs for each group is recommended). We suggest a screening method for dropping the less important KPIs in each group by comparing their weights with a minimum significant weight (*MSW*) as defined below:

$$MSW = \frac{1}{2x} \tag{1}$$

where 1 indicates that the KPIs are weighed between 0 and 1, and the sum of weights in each group should be equal to 1. The x is the number of indicators in the relevant group. The constant '2' is an empirical factor recommended based on our experience. We will have a *MSW* for each group. KPIs with weights less than relevant *MSWs* will be dropped from their groups, and the final leading and lagging KPI sets will be determined. It should be noted that KPIs that are obligated by regulations or recommended by authorities can be added to these groups as exceptions.

When the final KPIs of each group are determined, once again the AHP method is used to weigh the KPIs by pairwise comparison. The leading and lagging KPI groups should be weighed separately while the sum of weights in each group should be equal to 1. In the next step, KPIs are scored in a Likert scale form from 1 (very weak performance) to 5 (excellent performance) (Hinkin 1995). A minimum acceptable level (*MAL*) is also determined for each KPI. A *MAL* value for a KPI is actually the lowest acceptable limit for the performance of an organization in that KPI, which can be based on stakeholders' expectations. The *MAL* values are used for the evaluation of performance of each KPI. The ratings of KPIs and their minimum acceptable levels are the most critical steps in the proposed framework. The final result of performance measurement strongly depends on the conduct of this process, which is undertaken by an expert team. Therefore, the knowledge and experience of the expert team should not be underestimated.

Stage III: quantification of safety performance

Based on the results of the previous stages, in stage III, the required data for measuring KPIs should be gathered and recorded by companies or organizations. It is important to define two basic time intervals for assessing KPIs. First, the team decides on the time between two subsequent measurements of KPIs. Second, the team determines the time interval at which the measured KPIs are used to calculate the safety performance index. It should be noted that the index is based on the aggregation of a range of indicators or KPIs.

In this chapter, the KPI measurement period and the safety index measurement period were defined as one and five years, respectively. This means that the required data for the calculation of the selected KPIs should be gathered continually in order to determine their values each year. By repeating this process, at the end of fifth year, we will have enough data to calculate the first five-year safety index and determine safety performance. After the fifth year, we will have a new index for each year based on the most recent five-year data.

When the five-year KPI data have been recorded, it is time to calculate the safety index that reflects the safety performance. For this quantification process, a proper mathematical model is needed. In our model, four factors are calculated from each KPI value which are combined to calculate the safety index. The four factors are: Missing KPI Factor (*MKF*), Deviation Frequency Factor (*DFF*), Deviation Severity Factor (*DSF*) and Reward Factor (*RF*).

The Missing KPI Factor (*MKF*) is intended to punish the organizations for situations where a KPI is not calculated in the planned time, for any reason. It will be calculated as below:

$$MKF = \begin{cases} 0; & \text{if KPI is measured} \\ 1; & \text{if KPI is not measured} \end{cases} \tag{2}$$

The Deviation frequency factor (DFF) is defined to verify whether the KPI status is acceptable or not. For this reason, the value of the KPI will be compared with its 'minimum acceptable level (MAL)', as defined in stage II. The DFF will be:

$$DFF = \begin{cases} 0; & \text{if KPI value is greater than MAL} \\ 1; & \text{if KPI value is less than MAL} \end{cases} \tag{3}$$

The sum of all DFF values in a five-year period will be an indication of the frequency of MAL violation by KPIs. Furthermore, by using the KPI weights (defined in stage II), the model can account for differences in the violation of different KPIs, in the final equations (equations 5 and 6).

The Deviation Severity Factor (DSF) determines the magnitude of deviation from MAL, if the KPI value is less than MAL. It will be calculated as below:

$$DSF = \begin{cases} 0; & \text{if KPI value is greater than MAL} \\ MAL / KPI\,value; & \text{if KPI value is less than MAL} \end{cases} \tag{4}$$

These three factors (i.e. MKF, DFF and DSF) are the components of 'deviation from ideal performance'. These factors can be combined to calculate the amount of deviation in leading and lagging performance. We defined a Lead Deviation Index (DI_{lead}) and a Lag Deviation Index (DI_{lag}) which can be calculated from the leading and lagging KPIs as follows (formulas 5 and 6):

$$DI_{lead} = \left(\left(\frac{\sum_{i=1}^{n}\sum_{j=1}^{m}\left(MKF_{ij} \times C_1\right)}{(n \times m)} \right)^2 + \left(\frac{\sum_{i=1}^{n}\sum_{j=1}^{m} DFF_{ij} \times w_j \times m \times C_2}{(n \times m) - \sum_{i=1}^{n}\sum_{j=1}^{m}\left(MKF_{ij}\right)} \right)^2 + \left(\frac{\sum_{i=1}^{n}\sum_{j=1}^{m}\left(DSF_{ij} \times C_3\right)}{(n \times m) - \sum_{i=1}^{n}\sum_{j=1}^{m}\left(MKF_{ij}\right)} \right)^2 \right)^{1/2} \tag{5}$$

$$DI_{lag} = \left(\left(\frac{\sum_{i=1}^{n}\sum_{j=1}^{k}\left(MKF_{ij} \times C_1\right)}{(n \times k)} \right)^2 + \left(\frac{\sum_{i=1}^{n}\sum_{j=1}^{k} DFF_{ij} \times w_j \times k \times C_2}{(n \times k) - \sum_{i=1}^{n}\sum_{j=1}^{k}\left(MKF_{ij}\right)} \right)^2 + \left(\frac{\sum_{i=1}^{n}\sum_{j=1}^{k}\left(DSF_{ij} \times C_3\right)}{(n \times k) - \sum_{i=1}^{n}\sum_{j=1}^{k}\left(MKF_{ij}\right)} \right)^2 \right)^{1/2} \tag{6}$$

In these equations, n is the index measurement period in years (recommended to be 5), m is the number of selected lead KPIs and k is the number of selected lagging KPIs. w_j is the weight of KPI number j ($0 < w_j < 1$), determined in stage II. C_1 is the *MKF* correction coefficient, considered to adjust the effect of missing KPIs, C_2 is the *DFF* correction coefficient, considered to adjust the effect of deviation frequency, and C_3 is the *DSF* correction coefficient responsible for correcting the effect of deviation magnitude.

For a better understanding of equations 5 or 6, *MKF*, *DFF* and *DSF* may be considered as a matrix with n rows and m (or k) columns.

Apart from the negative aspects of performance (i.e., deviations), the methodology proposes a reward factor (*RF*) that reflects the positive aspects of performance, for situations where the performance in one KPI is better than its determined *MAL*. Hence, *RF* was defined as follows:

$$RF = \begin{cases} 0; & \textit{if KPI value is less than MAL} \\ \textit{KPI value/MAL}; & \textit{if KPI value is greater than MAL} \end{cases} \tag{7}$$

By averaging the *RF* values and correcting their effect, the positive reward for lead and lag performance can be calculated using equations 8 and 9:

$$Reward_{lead} = \frac{\sum_{i=1}^{n}\sum_{j=1}^{m} AF_{ij} \times C_4}{n \times m} \tag{8}$$

$$Reward_{lag} = \frac{\sum_{i=1}^{n}\sum_{j=1}^{k} AF_{ij} \times C_4}{n \times k} \tag{9}$$

In these equations, C_4 is the *RF* correction factor considered to adjust the effect of positive performance in the final performance indices.

The recommended values for the constants of the described mathematical model are: $C_1 = 14$, $C_2 = 6$, $C_3 = 5$ and $C_4 = 2$. Based on these values, the formulas for Lead Safety Performance Index (SPI_{lead}) and Lag Safety Performance Index (SPI_{lag}) were developed as:

$$SPI_{lead} = 22 + Reward_{lead} - DI_{lead} \tag{10}$$

$$SPI_{lag} = 22 + Reward_{lag} - DI_{lag} \tag{11}$$

These formulas are based on the renowned 'reward and punishment' policy (Nuttin & Greenwald 1968), which considers punishment for weak performance and reward for good performance. With any probable change in the model constants (C_1 to C_4), these two formulas (equations 10 and 11) should be revised to change their constant, 22.

From the results of equations 10 and 11, the overall Safety Performance Index (*SPI*) can be calculated by combining the lead and lag performance indices, as below:

$$SPI = \frac{1}{\sqrt{2}}\left(SPI_{lead}{}^{2} + SPI_{lag}{}^{2}\right)^{1/2} \tag{12}$$

In equations 5, 6 and 12, we used an analogy to Pythagoras' theorem (Collignon 1906) in vector algebra which is used for the calculation of the magnitude of vectors from their components. In practice, with the assigned constants ($C_1 = 14$, $C_2 = 6$, $C_3 = 5$ and $C_4 = 2$) and by a qualitative survey, this model indicated acceptable sensitivity to changes in KPI values. It is clear that the sensitivity of this model is dependent on the values of the constants used.

It should be noted that the constant values in the above formulas have been determined empirically based on the team experience and by a 'trial and error' process. After developing a spreadsheet program, the model was examined with different constants for different hypothetical KPI sets. Finally, the best set of constants have been selected to balance the effect of different factors (*MKF*, *DFF*, *DSF* and *RF*).

The final criteria for the evaluation of SPI_{lead}, SPI_{lag} and *SPI* were determined qualitatively by expert opinion (see the guideline in Table 7.2). The boundaries in this table were defined by calculating the performance indexes for different hypothetical KPI sets and by comparing the results with our expert team judgement. These criteria are used to judge the performance, after quantification process. This is similar to a fuzzification process, followed by defuzzifications and quantifications in the previous steps, which helps the users to comprehend the meaning of the final result. It is clear that in the case of any change in model parameters or constants, Table 7.2 may need revision.

In our framework, the *MAL*s are expected to be updated to be stricter over time, according to a continual improvement policy of organizations, but for

Table 7.2 The guideline for the evaluation of performance indices (for $C_1 = 14$, $C_2 = 6$, $C_3 = 5$ and $C_4 = 2$)

Class	Very poor	Poor	Relatively good	Good	Excellent
Index	< 10	≥ 10 & <15	≥ 15 & <20	≥ 20 & < 23	≥ 23

effective comparisons and benchmarks, same KPIs with same weights and *MALs* should be used.

The proposed framework covers the whole range of performance measurement, from 0 (identification of KPIs) to 100 (evaluation of aggregated index). It can be easily used in any kind of organization or company, regardless of its size, structure and activity. If same KPIs with same parameters are used, the performances can be compared. In measuring safety performance, there may be a need to measure the performance in a special field, such as personal safety. Similarly, an organization may want to compare separately the performance of different elements (human factor, equipment, management, etc.). On the other hand, it may be needed to measure the performance of the wider system, such as an Integrated Management System (IMS). The proposed framework can be easily modified for such narrow or extended applications. In the proposed methodology, the long-term performance index (e.g. five-year index) can be calculated from the values of different KPIs recorded more frequently (e.g. each one year). This is important because in the safety domain, the KPIs should be recorded continually in short-term intervals, but the performance can be judged in relatively longer intervals. It should be noted that although the model is designed to measure safety performance in the long term (e.g. for five years), it has no limitations to calculate the short-term performance (e.g. for one year or two years).

The proposed model uses the results of risk assessment. In case that there is no previous risk assessment study, it is suggested that the team performs a comprehensive baseline risk assessment in order to be facilitated in the selection of proper KPIs. On the other hand, the result of the risk assessment (e.g. the number of red risks) can be used as leading KPIs, or as 'inputs' to determine the safety performance status.

The developed model contains a *MKF* parameter that is defined to lower the performance index for the cases that a KPI is not measured for some reasons. The model enforces continual measurement of all KPIs and punishes the user for possible negligence. A correction coefficient is also defined to adjust *MKF* effects. In addition to *MKF*, two of other elements (i.e. *DFF and DSF*) can be used to compute the deviation frequency and the deviation severity, for the measured KPIs. If we consider 'deviation of a KPI from its *MAL*' as a hazard, then a combination of deviation frequency and severity may be an indication of the deviation risk. In this sense, the model uses a strong philosophy, consistent with the risk concept. The model also uses a reward factor (*RF*) to account for cases where the performance in one KPI is over *MAL*. Altogether, this model is based on the renowned 'reward and punishment' policy.

The model follows a simple and straightforward process to calculate performance indices. The output is an aggregated single number which can be used for effective communication with non-expert personnel, stakeholders, public and senior management. The model combines simple qualitative and quantitative processes for quality measurement and it can easily be learnt and used by safety staff. Initially, the lead and lag performance are calculated separately, but then they are combined to compute a safety performance index. Finally, the overall performance can be evaluated using the selected criteria or the guidelines in Table 7.2. This general framework can be considered as a basis for publishing an international ISO standard for measuring safety performance. The continual improvement proposed by ISO standards can be achieved with this framework by continual measurement of performance, and update of *MAL* values as needed. The continual measurement of performance within a certain period (one year, two years, three years, etc.) is easily accommodated in the presented simple framework.

The defined mathematical formulation, which is analogous to vector magnitude relation in vector algebra, has an acceptable sensitivity with respect to input KPI values. The four coefficients of the model can be used to customize and adjust the sensitivity of it. In addition, the model is not sensitive to the number of selected KPIs and the index calculation period. This is a very useful feature that increases flexibility of the model which can be customized according to the needs of different users.

A case study was undertaken by the author to compare the performance of three oil and gas drilling companies in Iran (Amir-Heidari et al. 2017). Besides that work, a useful spreadsheet program was developed which can be customized by users for their special purposes. An example of a spreadsheet workbook was also presented in that work. The programs can be downloaded from the Elsevier Web site (Amir-Heidari et al. 2017).

Conclusions

In this chapter, a logical selection and aggregation methodology was developed for generating a single number (or index) that reflects safety performance. This is very helpful for communicating the status of safety performance to non-expert personnel, stakeholders, senior management and the public. This simple and logical structure with a strong philosophical and mathematical base did not exist in the literature. Therefore, the three-stage framework, which is based on a 'reward and punishment' policy, can be used by many organizations or companies in measuring safety performance.

In the proposed framework, the aggregation result of stage III (lead, lag and overall indices) is suitable for general communication, although for professional purposes, safety managers of organizations should scrutinize the detailed results of the evaluation of all KPIs to discover any weaknesses and plan for future improvements. This framework can become a base for developing an international ISO standard for safety performance monitoring.

References

Amir-Heidari, P, Maknoon, R, Taheri, B & Bazyari, M 2017, 'A new framework for HSE performance measurement and monitoring', *Safety Science*, vol. 100, pp. 157–67, https://doi.org/10.1016/j.ssci.2016.11.001

Antao, P, Calderon, M, Puig, M, Michail, A, Wooldridge, C & Darbra, RM 2016, 'Identification of Occupational Health, Safety, Security (OHSS) and Environmental Performance Indicators in port areas', *Safety Science*, vol. 85, pp. 266–75. doi:10.1016/j.ssci.2015.12.031

Collignon, ME 1906, 'Pythagoras theorem', *Proceedings of the Edinburgh Mathematical Society*, vol. 25, pp. 91–4, http://dx.doi.org/10.1017/S0013091500033605

Dalkey, N 1969, 'An experimental study of group opinion: The Delphi method', *Futures*, vol. 1, no. 5, pp. 408–26. doi:10.1016/S0016-3287(69)80025-X

Haas, EJ & Yorio, P 2016, 'Exploring the state of health and safety management system performance measurement in mining organizations', *Safety Science*, vol. 83, pp. 48–58. doi:10.1016/j.ssci.2015.11.009

Hale, A 2009, 'Why safety performance indicators?', *Safety Science*, vol. 47, no. 4, pp. 479–80. doi:10.1016/j.ssci.2008.07.018

Hassan, J & Khan, F 2012, 'Risk-based asset integrity indicators', *Journal of Loss Prevention in the Process Industries*, vol. 25, pp. 544–54. doi:10.1016/j.jlp.2011.12.011

Hendershot, D 2007, 'Measuring process safety performance', *Journal of Chemical Health & Safety*, pp. 47–8. doi:10.1016/j.jchas.2007.07.009

Hinkin, TR 1995, 'A review of scale development practices in the study of organizations', *Journal of Management*, vol. 21, no. 5, pp. 967–88. doi:10.1016/0149-2063(95)90050-0

Hopkins, A 2007, 'Thinking about process safety indicators', *Working Paper 53*, National Research Centre for OHS Regulation, Australian National University.

Juglaret, F, Rallo, J, Textoris, R, Guarnieri, F & Garbolino, E 2011, 'New balanced scorecard leading indicators to monitor performance variability in OHS management systems', *Proceedings of the fourth Resilience Engineering Symposium*, pp. 121–7. Sophia Antipolis, France.

Nuttin, J & Greenwald, AG 1968, 'Reward and punishment in human learning: Elements of behavior theory', Elsevier Inc., ISBN: 978-1-4831-9723-4.

Podgorski, D 2015, 'Measuring operational performance of OSH management system: A demonstration of AHP-based selection of leading key performance indicators', *Safety Science*, vol. 73, pp. 146–66. doi:10.1016/j.ssci.2014.11.018

Rogers, W & Pasman, H 2013, 'How can we use the information provided by process safety performance indicators? Possibilities and limitations', *Journal of Loss Prevention in the Process Industries*, vol. 30, pp. 1–10. doi:10.1016/j.jlp.2013.06.001

Saaty, RW 1987, 'The analytic hierarchy process: What it is and how it is used', *Mathematical Modelling*, vol. 9, no. 3–5, pp. 161–76. doi:10.1016/0270-0255(87)90473-8

Swuste, P, Theunissen, J, Schmitz, P, Reniers, G & Blokland, P 2016, 'Process safety indicators: A review of literature', *Journal of Loss Prevention in the Process Industries*, vol. 40, pp. 162–73. doi:10.1016/j.jlp.2015.12.020

Wu, X, Liu, Q, Zhang, L, Skibniewski, MJ & Wang, Y 2015, 'Prospective safety performance evaluation on construction sites', *Accident Analysis and Prevention*, vol. 78, pp. 58–72. doi:10.1016/j.aap.2015.02.003

Safety by design

Design of inland container terminals with dangerous goods

Gemma Dolores Molero, Francisco Enrique Santarremigia, Pablo Aragonés-Beltrán and Juan-Pascual Pastor-Ferrando

Introduction

Need

Sales of chemical products have increased a 14 percent in 2015, from €3,100 billion in 2014 to €3,534 billion in 2015. Chemical production is mainly driven by China, with total sales of €1,409 billion in 2015. The EU chemical industry is the second in the ranking, with an income of €615 billion in 2015 (17.4%), followed by NAFTA (North American Free Trade Agreement) with €583 billion sales (CEFIC 2016).

On the one hand, the increase in the number of containers with hazard goods, without associated infrastructure investment, is producing an increase of traffic jams, ports' yard congestion, road congestion around port gates, and higher environmental impact in ports´ surrounding areas (European Commission 2013a). Moreover, one-fifth of the containers entering or leaving Europe are managed in just three ports: Rotterdam, Hamburg and Antwerp. This imbalance between ports' performance results in congestion and extra costs for shippers, transport operators and consumers (European Commission 2013b). These problems can be solved by the development of multimodal *inland terminals*, also known as *dry ports*, which are directly connected to seaports and where customers manage their goods as they would do in the seaport. The use of this type of facility will decrease traffic congestion in gateway ports.

On the other hand, the storage of dangerous goods is regulated for warehouses but not for seaports or inland terminals. The characteristics of dangerous goods need special attention for their handling, storage and management. However, to date, there are not any guidelines or standards for the design of inland terminals with dangerous goods (DGs), what would help in a safer management of the terminal according with the pillar "Commitment in action" of Total Safety Management (TSM).

The main drawbacks of the storage of containers with dangerous goods in ports are:

1 The maritime and inland (road and rail) transport of contain with dangerous goods is increasing every year. Road freight transport of dangerous goods in Europe moved around 82 billion ton-kilometres in 2015, showing an 8.8% increase compared with 2014.

(Eurostat 2016)

2 There is not a unified and common European regulation for the storage of containers with dangerous goods in ports.
3 Storage limitations in seaports. Currently, national regulations in Europe allow a container with DGs to stay only a limited time in the port yard.
4 Multiple design criteria interfere in the design of inland terminals.
5 There is a lack of design guidelines for inland terminals with dangerous goods.

To overcome these drawbacks the *main objective* of this chapter is to develop a methodology from a TSM perspective for the design of terminals with dangerous goods, providing companies and designers involved in this subject with a tool that allows them to identify, analyse and weigh the different criteria involved in the design in a structured way.

The *specific objectives* addressed in this chapter are:

1 To explain the concepts related with inland terminals and dangerous goods (e.g. risks, storage concerns)
2 To carry out an exhaustive analysis of current state of the art to identify criteria involved in the design of inland terminals with dangerous goods, and to classify them in different study disciplines

3 To form an expert panel for the identification of extra criteria and for criteria weighting in order to consider stakeholder´s needs
4 To define a model based on a Multicriteria Decision Analysis (MCDA) method, the Analytic Hierarchy Process (AHP). This model converts a complex problem into various sub-problems that can be analysed separately in order to simplify the procedure. The analysed factors can be quantitative and/or qualitative. AHP allows the classification and prioritization of factors based on experts opinions.

Design of inland container terminals with dangerous goods towards Total Safety Management

Currently, there is not any procedure for the management and storage of dangerous goods in terminals of containers from a Total Safety Management perspective. Within the term "dangerous goods", we can find different classes of goods (acid, base, flammable, corrosive, toxic, etc.), which have different product compatibilities and requirements for storage conditions. The increase of chemical sales and their storage in terminals in an inappropriate way without a good regulation makes them very dangerous, producing accidents as the one in Tianjin (China) in 2015, with 173 deaths and damages in a radius of two kilometres.

Therefore, an appropriate and safe management of a port or an inland terminal of containers is needed, starting from a proper design of the layout of terminals from different points of view (e.g. equipment performance, safety, security against intruders, Information and Communication Technologies [ICT], environment and Business Intelligence [BI]) and taking into account safety criteria and stakeholders' needs.

State of the art

Dangerous goods and hazardous materials

Definition

Dangerous goods (also known as hazardous materials or its abbreviation HAZMAT in the United States) are any material, product, or packaged or bulk substance that can constitute a risk to people, animals or the environment if they are not handled properly during their use or in their transport. They are often subjected to chemical regulations and must be transported according to a set of official rules.

Regulations and classification

It is important to consider a proper classification of dangerous goods for the correct functioning of the supply chain.

There are three main actors in the classification of dangerous goods:

1. The *UN Manual of Tests and Criteria* that establishes classification procedures, test methods and criteria related to each type of dangerous good. This manual determines the dangerousness of goods and their classification into different packaging groups (I, II or III).
2. The *UN Orange Book* also called *UN Recommendations on the Transport of Dangerous Goods – Model Regulations* (United Nations 2015) that classifies and assigns all dangerous goods into a UN number and to a shipping name based on their risk and hazard classification.
3. *Transport codes for each mode of transport:* ADR (European Agreement concerning the International Carriage of Dangerous Goods by Road), IMDG (International Maritime Dangerous Goods Code), RID (International Carriage of dangerous goods by Rail), DGR-IATA (Dangerous Goods Regulation from the International Air Transport Association), ADN (The European Agreement concerning the International Carriage of Dangerous Goods by Inland Waterways).

The European regulations related to the transport and storage of DGs (see Figure 8.1) are:

- Directive 2008/68/EC for the inland transport of dangerous goods. This Directive establishes a common regime for all aspects of the inland transport of dangerous goods, by road, rail and inland waterway.
- Regulation 1272/2008/EC on classification, labelling and packaging of substances and mixtures, also called "CLP".

Dangerous goods storage in ports

Chemical products storage in ports are regulated by national laws around Europe. As an example, Spain applies the RD 145/1989 about the National

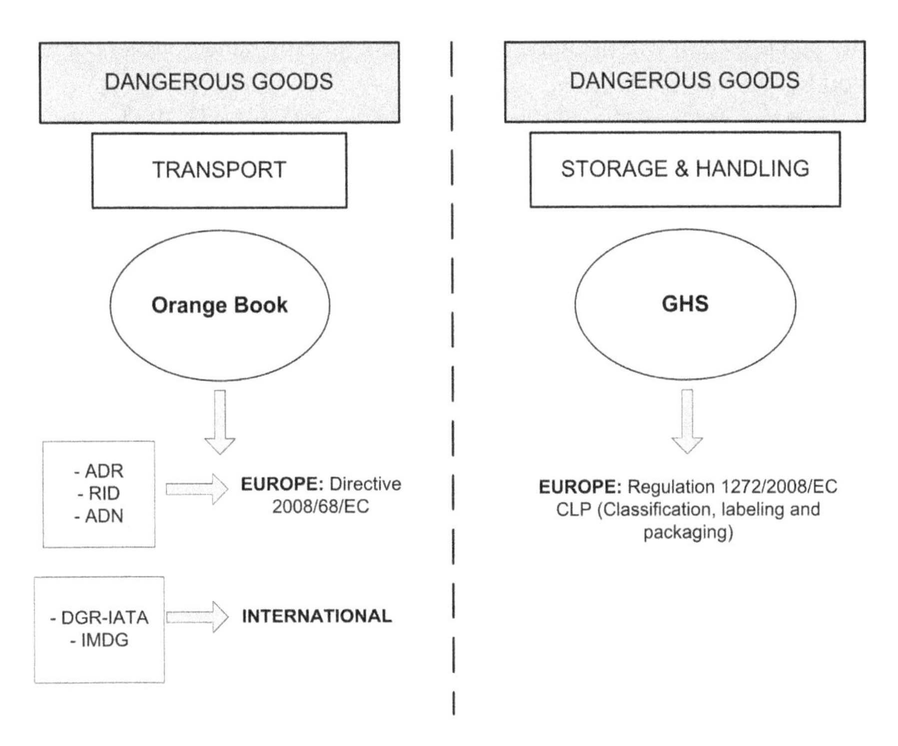

Figure 8.1 Summary of regulations and codes applied to dangerous goods and hazardous substances. GHS: Globally Harmonized System of Classification and Labelling of Chemicals

Regulation of Admission, Handling and Storage of Dangerous Goods in the Ports, and Law 48/2003, modified by Law 33/2010. These regulations create incompatibilities between different chemical products that have to be considered for storage purposes (Sanz Albert 2014).

There are other national legislations dealing with the segregation of dangerous products in the industry, for example: the RD 379/2001 in Spain, the Health and Safety Guidance 71 (HSG-71) in the United Kingdom, or the GefStoffV Ordinance, the VCI (German Chemicals Industry Association) guideline for the joint storage of chemicals and the Technical rule (TRGS 510) in Germany.

Nowadays, temporary storage restrictions in ports are not clear due to the lack of a joint European regulation. Some national regulations have established that a container with any kind of hazard can stay in the port for only a limited period of time (e.g. eight days). Storage in the port yard for more days is forbidden unless there is a specific authorization. The duration of the authorization will depend on the amount and the dangerousness of the goods.

From these reasons, it is necessary to design inland terminals, also known as dry ports, where containers with dangerous goods can be stored for more time, allowing more flexibility to the costumers and minimizing any adverse impact for the environment. This serves the principles of total safety management.

Dry ports

Dry ports, also known as inland terminals, inland ports or inland freight terminals, are intermodal terminal facilities in the hinterland, directly connected to seaports and where customers manage their goods as they would do in the seaport (Roso 2007).

The development, organization and integration of inland ports in the different areas of the world depend mainly on regional characteristics. Differences between dry ports and their integration with seaports depend on two main factors: the characteristics of the port (e.g. geographical distribution, modal availability, regulations, logistics network, etc.) and the activities developed in the port (e.g. volume of the cargo, flow frequency, type of good).

Main strengths of the use of dry ports are:

- Lower traffic congestion and environmental impact in areas near the seaports due to the reduction of the number of the transportation connections to/from the seaports
- Fewer overcrowded ports
- Lower transportation costs
- Intermodal and co-modal transport solutions. The use of rail transport, which is faster than the use of trucks, and good planned logistic networks will improve the seaport´s capacity without physical expansion of the port.
- Development of safer ports

Inland terminals design factors: literature review

The design of an inland terminal where dangerous goods are going to be managed is not a problem with an immediate solution. In the design of a terminal multiple factors need to be considered: equipment, security against intruders, Information and Communication Technologies (ICT), etc. In addition, ports handling dangerous goods should take into account extra considerations from the safety point of view in order to be secure and avoid any dangerous event. Chemical spills can harm the workers or the equipment, and a mix of incompatible chemical products can produce a fast and incontrollable chemical reaction that could endanger the entire terminal.

The different design factors found in the literature can be classified into the following categories: equipment, ICT and Business Intelligence (BI), safety and security and environmental criteria (Molero et al. 2017).

Most methodologies applied in this field include: cost models, multicriteria decision methods, risk assessments, mathematical models and simulations. Costs models have focused on the evaluation of equipment performance and the determination of the storage space (Kim & Kim 2002). Moreover, Saanen, van Meel, and Verbraeck (2003) used cost models together with simulation technologies for the evaluation of the operational productivity of the design and security of the new generation of automatized systems (ICT) for container terminals. Regarding the environmental area of study, Linkov and Ramadan (2004) developed comparative risk assessments in different case studies to help in Environmental Decision Making. Moreover, mathematical models have been used to optimize the equipment performance (Kaysi & Nehme 2016).

Multicriteria Decision making Methods (MCDM) have been widely used to help in the decision making process. For example, Yang, Tai and Chiu (2014) determined the importance of various factors which influence the infrastructure used for the transport of containers in port terminals (e.g. operational cost, operational efficiency, or transshipment time). In addition, Liu, Jula and Ioannou (2002) utilized an MCDM to determine the optimum number of automated guided vehicles (AGV) using ICT and BI indicators. Other type of criteria evaluated by MCDM was environmental care. Yang (2015) used MCDM to analyze green container terminals in order to do a ranking between six different ports based on their environmental sustainability.

Safe design of seaport/inland container terminals

Methodology

Usually the methods employed for the design of terminals of containers have focused only on a few issues e.g. port capacity, number of imports and exports or berth capacity. However, the analysis of the state of the art made it clear that, for an appropriate and safe design of container terminals, a holistic approach is needed in order to consider a whole range of factors. Special attention should be paid to the importance of each factor or the criteria influencing the design of a container terminal fulfilling requirements for safety, security, operational and environmental issues. Therefore, a proper study needs the involvement of experts in the field and multicriteria tools to help in the decision making process. Multicriteria Decision Making Methods are very useful techniques when different criteria are considered in order to develop a project. The main objective of these techniques is to evaluate and choose between alternatives based on the evaluation of different criteria by a group of experts or work group.

In order to obtain a ranking with the most important factors to be considered in the design of container terminals, the following methodology (Molero 2016) can be used (Figure 8.2).

Analysis phase

The analysis phase involves two main steps: the definition of the problem of study and the study of the criteria involved for a proper analysis of the

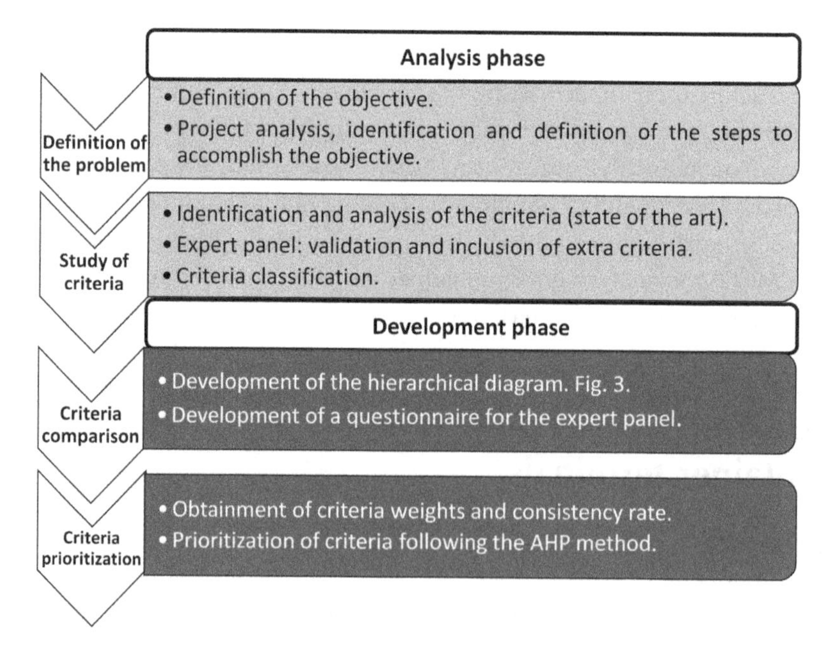

Figure 8.2 Scheme of the MCDM steps followed for the prioritization of criteria for the design of safe terminals of containers

problem. For the definition of the problem, collaborations with companies involved in management of seaport and inland terminals are very useful for a good definition and development of the work.

For the study of the criteria involved in container terminals with dangerous goods, an exhaustive analysis of the state of the art was done. Identified criteria were validated by an expert panel created according with the Stakeholder theory (Reynolds, Schultz & Hekman 2006). Meetings with other expert members were carried out in several scientific forums (COP21 Paris Climat 2015; TraConference Paris 2014) and led to the validation of criteria and to the identification of new criteria (TraConference Warsaw 2016).

Development

The development of the methodology can be divided in two main steps: the criteria comparison and the criteria prioritization. Both steps follow the multicriteria decision making methodology called Analytic Hierarchy Process (AHP) (Saaty 1987; Saaty & Kułakowski 2016). In order to obtain the weights of the criteria, first, a hierarchical diagram was developed for grouping the different criteria at different levels (see Figure 8.3). Afterwards, a

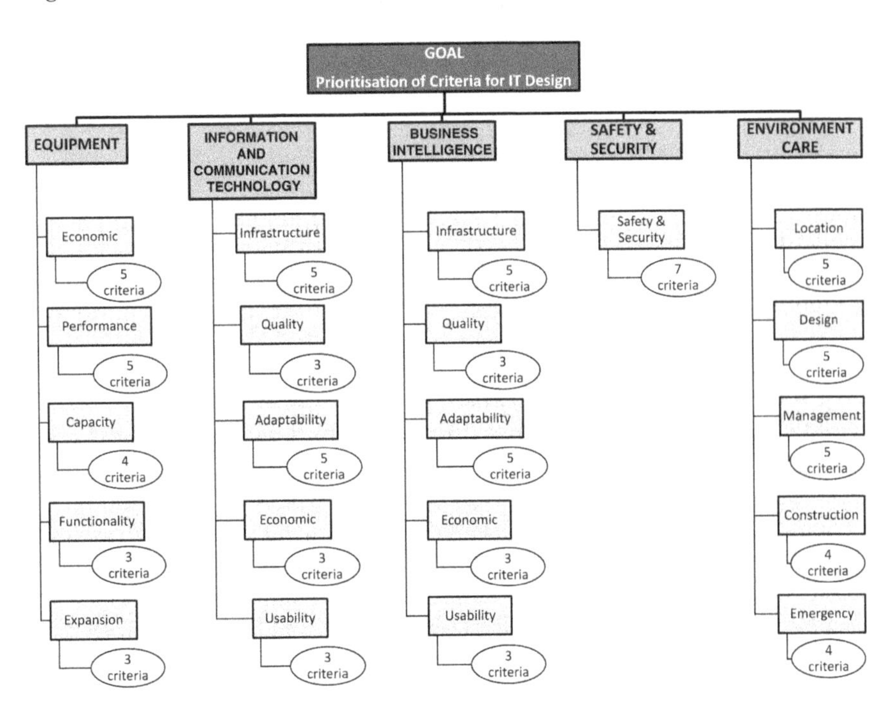

Figure 8.3 Hierarchical diagram showing three levels of criteria for IT design

questionnaire was prepared for the expert panel for the comparison of criteria from the second and the third level. In these questionnaires criteria are compared by pairs. Results of questionnaires and the preference of a criterion over another have been translated into the scale of Saaty (Saaty 1987):

1 – Similarly preferred. Both elements are equally preferred.
3 – Moderately preferred. The element in the row is slightly preferred.
5 – Strongly preferred. The element in the row is strongly preferred.
7 – Very strongly preferred. The element in the row is very strongly preferred.
9 – Extremely preferred. The element in the row is extremely preferred.

Criteria 2, 4, 6 and 8 are intermediate values.

Consensus between all members of the expert panel has been achieved following the DELPHI method (Linstone & Turoff 2002).

The data collected from the expert panel constituted matrices of comparison and local and global weights of criteria for the different levels and their prioritization has been obtained following the AHP methodology (Saaty 2016) using *SuperDecisions* software.

For the calculation of normalized local weights for second level criteria (w_{cj}) the following procedure was used:

1 The n criteria in the same level were compared in pairs using Saaty's scale. These values form, for each level, a comparison matrix A:

$$A = \begin{bmatrix} 1 & a_{12} & \cdots & a_{1n} \\ a_{21} & 1 & \cdots & a_{2n} \\ \vdots & \vdots & \ddots & \vdots \\ a_{n1} & a_{n2} & \cdots & 1 \end{bmatrix}, \; where\; a_{ji} = 1/a_{ij} \qquad i,j = 1,\ldots,n \tag{1}$$

Consistency ratio (CR) of matrix A was used to check inconsistencies. Matrices with a CR of above 0.1 were rejected.

(Saaty 1987)

2 The local normalized weight of each criterion $w = (w_{c1}, w_{c2}, \ldots, w_{cj}, \ldots, w_{cn})$ is the principal eigenvector of the pairwise comparisons matrix A, which was calculated by raising this matrix to a sufficiently large power:

$$q_{jl}^{k} = \lim_{k \to \infty} A^{k} \tag{3}$$

Then summing over the rows and normalizing, we obtained the local normalized weight in the kth power:

$$w_{cj}^{k} = \frac{\sum_{l=1}^{n} q_{jl}^{k}}{\sum_{j=1}^{n}\sum_{l=1}^{n} q_{jl}^{k}} \quad (4)$$

The process was stopped when the difference between w_{cj}^{k} obtained at the kth power and w_{cj}^{k+1} obtained at the $(k + 1)$th power was less than 10^{-4}.

3) The same process was applied to calculate the local normalized weight for each criterion of third level w_{ck}.
4) Global normalized weights (W_{CG_k}) are calculated using (5).

$$W_{CG_k} = w_{ci} \cdot w_{cj} \cdot w_{ck} \quad (5)$$

Where w_{ci} is the local normalized weight of the first level criterion, w_{cj} the normalized weight for the second level criterion and w_{ck} the normalized weight of the third level criterion.

All this process can help us conclude with a hierarchy of criteria related with the design of inland terminals.

Results and productivity challenges

After the application of the described methodology; the global normalized weights were obtained for the different criteria employed. The full list of criteria with their definitions and the obtained weights can be seen in Molero et al. (2017).

The ranking of the third level criteria shows that the 40% of the criteria with the greatest weights suppose the 80% of the sum of the weights. Inside this group, "safety and security" is the most important type of criteria, supposing 23% of the weight, and "environmental care" supposes 17%. This highlights the need for a safety management of the terminal from the beginning, taking into special consideration safety issues in the design stage.

In addition, from a total of 88 criteria, the ten criteria with the highest weights sum the 45%. Two important criteria from the safety and security area included: the distance to the urban core and the danger level of the dangerous goods. Prevention of disasters is imperative when dangerous substances are present in a container terminal. Thus, the distance in kilometres from the terminal to the nearest population core, together with the hazard characteristics of the goods (toxicity, explosion, fire, etc.) and the risks of thermal radiation and pressure waves that may be produced during accidents are very important criteria in the design of terminals of containers with dangerous goods.

Another important parameter that has a high influence in the design of the terminal is the criterion ground cost from the equipment area. This parameter was considered in the quantification of the land cost measured in euros per square metre. Of course, the feasibility of a project necessarily involves profitability. Otherwise, we would be designing systems with little real applicability, so that the economic criteria have great relevance.

The availability of industrial floor area without the need to reassess rural lands is the third criterion in importance. It is a very beneficial criterion for project promoters as it can safeguard sustainable urban planning.

The *sensitivity analysis* allowed an analysis of how minor changes in decision makers' preferences can affect the hierarchy of criteria, and therefore the final decisions (Al-Harbi 2001). Small variations in each w_i (first level criteria weights), from 0.2 to 0.3, maintaining the other first level criteria at 0.175, were done and variations in the ranking of third level criteria were studied:

i) The sensitivity analysis showed that for the ten most important criteria (for the "Equipment", "Safety and Security" and "Environment care" areas) small changes were produced in the ranking of third level local weights.

(see Figure 8.4)

ii) The sensibility analysis also showed that criteria representing the 80% of the sum of the global weights are almost the same in all the cases simulated in this study.

(Molero et al. 2017)

Therefore, it can be concluded that the methodology and results are reliable and robust.

The productivity will be higher if the criteria weights obtained by the decision making process are considered in the preliminary design phase of inland terminals storing dangerous goods. These criteria have been obtained using a decision making model and following a planned sequence of decisions that is based on the preferences of all the stakeholders.

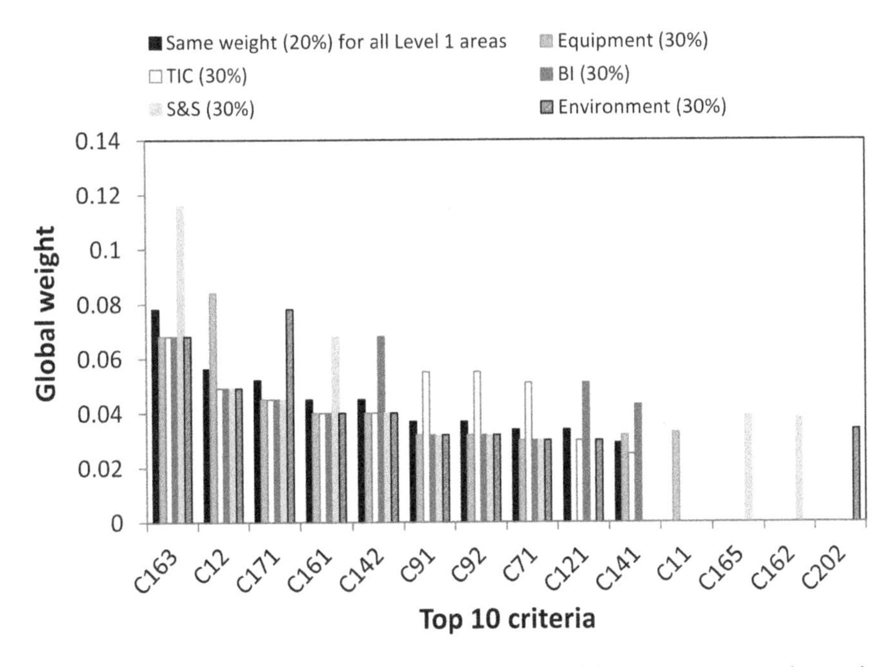

Figure 8.4 Comparison between the global weights of the top 10 criteria obtained in each of the six cases evaluated in the sensitivity analysis

Implementation in SMEs

Nowadays, the research group is working on further developments using the obtained criteria weights. Our recent studies have been focused on the identification and selection of layout alternatives, based on Container Handling Equipment (CHE) used in the yard, which best fit the characteristics and needs of each container terminal. This work has been presented in the International Congress on Transport Infrastructure and Systems (TIS) Roma 2017 (Santarremigia et al. 2017).

In this specific application, a GUI (Graphical User Interface) has been developed to bridge the gap between academia and the final users (most of them being SMEs). The tool is called CO_2 Multidecision Terminals,[1] and provides a prioritization of possible layout alternatives by means of the prioritization of CHE alternatives used in the yard due to the univocal relationship between CHE and the layout design. The final user of the tool can assign weights to criteria belonging to the first level (equipment, environmental and safety and security). Weights of second and third level criteria, which were established by the methodology through the expert panel according with the method described in paragraph "Methodology", are part of the data integrated in the tool. The tool allows the users to analyse the design of the layout of the terminal from a low carbon perspective (if the given weight for the environmental criteria is high) and compare it with the other possible scenarios obtained by the assignment of different weights to each of the three main areas of criteria.

Further developments should work on new dry port concepts based on efficiency, safety, security and environmental care criteria for a more sustainable transport of chemicals and dangerous goods.

Impact

The consideration in the design stage of all criteria identified and weighted in this work will conclude in more efficient, safer, more secure, resilient and environmental care inland terminals with dangerous goods. In order to assess the effectiveness of the selected approach or management strategy, it is needed to define a series of variables, called key performance indicators (KPIs), that have to be measured and compared in order to evaluate the overall performance of the implemented strategy.

Some examples of KPIs to follow up the performance of the implemented layout design can be seen in Table 8.1.

Table 8.1 Example of Key Performance Indicators to measure the performance of a dry port with dangerous goods

KPI	Definition or objective (measurement unit)
Consumption costs	Reduction of consumption costs (€/year).
Dangerous waste generation	Tons of dangerous waste generated (spillages, oils, contaminated filters, etc.) (tons/year).
Severity in the case of a serious accident (e.g. fire, spillage, explosion)/ terrorist attack	Number of persons likely to be affected in the case of accident (number of injured people and number of deaths) and monetary value of the damages; e.g. material losses and cost needed for the recovery of the affected environment (€).
Greenhouse gases emissions	Reduction of greenhouse gases emissions (tons CO_2 eq per unit of specific goods transported).

Conclusions

This chapter addresses the design of inland container terminals with dangerous goods (usually operated by SMEs) dealing with safety from a total management approach. Through the different sections of the chapter we have reached the following conclusions:

1 Dangerous goods are substances which can harm people or the environment, being subjected to regulations and a set of official rules and codes for their handling, transport and storage. The involvement of dangerous goods in an accident could produce very serious consequences with important human and material loses.
2 The classification of dangerous goods is handled by the *UN Orange Book*, the *UN Manual of Tests and Criteria* and the transport codes for each mode of transport.
3 Chemical Product Storage in ports is legislated by national laws. However, temporary storage restrictions in ports are not clear due to the lack of a joint European regulation.
4 The storage drawbacks, together with the increase in sales of dangerous goods and the subsequent increase in traffic jams and port congestion has led to the necessity of the creation of dry ports or inland terminals, most of them operated by private companies and SMEs.
5 Design factors of safe inland terminals with dangerous goods can be grouped in five main classes of criteria: equipment, ICT, Business Intelligence, safety and security and environmental criteria.
6 The AHP multicriteria decision making technique was used for the obtainment of weights for the identified criteria. These weights can be used by decision makers for the design of new dry ports or in the study of new investments from a TSM perspective. The sensitivity analysis showed that the obtained results are reliable and robust.
7 An application of this work has been presented by the application of some of these criteria (equipment, safety and security and environmental criteria) and their weights for the design of the layout of inland terminals with dangerous goods by the prioritization of CHE used in the yard of the terminal. CO_2 Multidecision Terminals is a tool marketed by an SME[2] bringing this layout designing tool to container terminals managers (most of them SMEs).

Notes

1 www.aitec-intl.com/en/project-IDi/22/co2-multidecisions-terminals/
2 AITEC (Valencia – Spain). www.aitec-intl.com

References

Al-Harbi, KMAS 2001, 'Application of the AHP in project management', *International Journal of Project Management*, vol. 19, no. 1, pp. 19–27.

CEFIC 2016, *Facts & Figures 2016 of the European chemical industry*, http://fr.zone-secure.net/13451/186036/#page=1

European Commission 2013a, *Europe's seaports 2030: Challenges ahead*, http://europa.eu/rapid/press-release_MEMO-13-448_en.htm

European Commission 2013b, *Commission proposes upgrade for 300 key seaports*, http://europa.eu/rapid/press-release_IP-13-451_en.htm

Eurostat 2016, *Road freight transport by type of goods-Eurostat*, http://ec.europa.eu/eurostat/statistics-explained/index.php/Road_freight_transport_by_type_of_goods

Kaysi, I & Nehme, N 2016, 'Optimal investment strategy in a container terminal: A game theoretic approach', *Maritime Economics & Logistics*, vol. 18, no. 3, pp. 250–63.

Kim, KH & Kim, HB 2002, 'The optimal sizing of the storage space and handling facilities for import containers', *Transportation Research Part B: Methodological*, vol. 36, no. 9, pp. 821–35.

Linkov, I & Ramadan, AB 2004, *Comparative risk assessment and environmental decision making*, www.springer.com/us/book/9781402018954

Linstone, HA & Turoff, M 2002, 'The Delphi method: Techniques and applications', *The Delphi Method: Techniques and Applications*, pp. 1–616, www.millennium-project.org/FRMv3_0/04-Delphi.pdf

Liu, C-I, Jula, H & Ioannou, PA 2002, 'Design, simulation, and evaluation of automated container terminals', *IEEE Transactions on Intelligent Transportation Systems*, vol. 3, no. 1, pp. 12–26.

Molero, GD 2016, 'Análisis de criterios de diseño básico de una terminal de contenedores de sustancias químicas peligrosas aplicando el proceso analítico jerárquico (AHP) (Analysis of criteria for the basic design of a container terminal of hazardous goods applying the Analytic Hierarchy Process (AHP)', *Universitat Politècnica de València*, pp. 195–244, http://dx.doi.org/10.4995/Thesis/10251/61773

Molero, GD, Santarremigia, FE, Aragonés-Beltrán, P & Pastor-Ferrando, J-P 2017, 'Total safety by design: Increased safety and operability of supply chain of inland terminals for containers with dangerous goods', *Safety Science*, vol. 100, pp. 168–182, http://dx.doi.org/10.1016/j.ssci.2016.10.007

Reynolds, SJ, Schultz, FC & Hekman, DR 2006, 'Stakeholder theory and managerial decision-making: Constraints and implications of balancing stakeholder interests', *Journal of Business Ethics*, vol. 64, no. 3, pp. 285–301.

Roso, V 2007, 'Evaluation of the dry port concept from an environmental perspective: A note', *Transportation Research Part D: Transport and Environment*, vol. 12, no. 7, pp. 523–7.

Saanen, Y, van Meel, J & Verbraeck, A 2003, 'The design and assessment of next generation automated container terminals', in *Simulation in industry-15th European simulation symposium* (pp. 577–84), SCS-European Publishing House, Germany.

Saaty, TL 1987, 'The analytic hierarchy process: What it is and how it is used', *Mathematical Modelling*, vol. 9, no. 3, pp. 161–76.

Saaty, TL 2016, 'The analytic hierarchy and analytic network processes for the measurement of intangible criteria and for decision-making', *International Series in Operations Research & Management Science.*, vol. 233, pp. 363–419.

Saaty, TL & Kułakowski, K 2016, *Axioms of the Analytic Hierarchy Process (AHP) and its generalization to dependence and feedback: The Analytic Network Process (ANP)*, pp. 1–12, http://arxiv.org/abs/1605.05777

Santarremigia, FE, Molero, GD, Esclapez, MD & Awad-Núñez, S 2017, 'Total management tool oriented to carbon footprint reduction in terminals of containers', in *Transport infrastructure and systems: Proceedings of the AIIT international congress on transport infrastructure and systems: Emerging technologies to enable smarter, greener and more efficient movement of people and goods around the world* (pp. 987–95), CRC Press, Taylor & Francis Group, LLC, Rome.

Sanz Albert, F 2014, 'Almacenamiento de productos químicos. Orientaciones para la identificación de los requisitos de seguridad en el almacenamiento de productos químicos peligrosos (Storage of chemical products: Guidance on the identification of safety requirements for the storage of hazardous chemicals)' (pp. 55–57), *Servicio de Ediciones y Publicaciones del INSHT*.

United Nations 2015, *Recommendations on the transport of dangerous goods: Model regulations*, United Nations Publication, New York and Geneva, www.unece.org/trans/danger/publi/unrec/rev19/19files_e.html

Yang, C-C, Tai, H-H & Chiu, W-H 2014, 'Factors influencing container carriers' use of coastal shipping', *Maritime Policy & Management*, vol. 41, no. 2, pp. 192–208.

Yang, YC 2015, 'Determinants of container terminal operation from a green port perspective', *International Journal of Shipping and Transport Logistics*, vol. 7, no. 3, p. 319, www.inderscience.com/link.php?id=69123

PART

4

Managing knowledge, skills and changes in the context of business strategy

Knowledge management in total safety for major hazards plants

Olga Aneziris, Zoe Nivolianitou, Myrto Konstantinidou, Emmanuel Plot and George Mavridis

Introduction

The aim of this chapter is to present the framework for Total Safety Management in the process industry by integrating safety and risk management approaches. As mentioned in Chapter 2, the cornerstone of this framework is the "Common Operational Picture" notion, which involves a useful synthesis of the unit risk assessment, with the intention to provide understandable information to the relevant stakeholders and decision makers. This framework was applied to the storage area of dichloropropene, a hazardous substance used for pesticides production. The Fault Tree and the bow-tie methods have been used for the risk assessment of various accidental releases from the storage tanks. Data, barrier information critical to safety and results of risk analysis are stored in an associated tool developed to incorporate the total safety management framework. This tool incorporates not only risk assessment methods such as Fault Trees and Bowties but also aspects of the plant safety management system, visualization tools for safety equipment, parts of the installation and risk zones in a 3D environment.

The need for knowledge management in total safety

Over the last 30 years various methodologies and tools have been developed playing a significant role in the safety assessment of plants involving toxic, flammable and explosive substances. Several state of the art reviews for risk assessment methodologies have appeared in the literature (Khan & Abbasi 1998; Tixier et al. 2002; Villa et al. 2016; Aneziris et al. 2014) for both qualitative and quantitative risk assessment methodologies. Qualitative methods include the HAZOP method, widely used in the chemical industry, which investigates deviations of all process variables in a plant together with their causes and consequences; the Failure Modes and Effects Analysis (FMEA), which can allow the evaluation of multiple failures, as reported by the Center of Process Safety (CCPS 2000); and the Master Logic method developed by Papazoglou and Aneziris (2003) which is a logic top down diagram and identifies initiating events leading to accidents. Quantitative probabilistic methods, include Fault Trees, Event Trees, Bowties and Bayesian networks, aiming at assessing risk and improving safety performance as proposed by several researchers (Villa et al. 2016; Ale et al. 2014; Khan and Abbasi 1998; Tixier et al. 2002; Papazoglou et al. 1992).

In recent years, standards for risk management, risk communication and risk mitigation have been developed, namely the ISO 31000:2009 and the NORSOK Z-013, 2010. The basic elements of risk management, as have been reported in these standards, are the following: (*a*) establishing the internal context of the organization (such as internal stakeholders, information flows, policies, resources and knowledge) and external parameters relevant to the environment in which the organization operates; (*b*) risk identification, analysis and evaluation; (*c*) risk treatment; (*d*) monitoring and reviewing of risk assessment; (e) application of risk; (*f*) communication and consultation with stakeholders and (*g*) monitoring and reviewing risk assessment so as to ensure continuous improvement in safety of the work environment and the personnel. The major steps for risk assessment as have been presented in these standards, but also by Papazoglou et al. (1992), are hazard identification, accident sequence modelling and quantification, consequence assessment and finally integration of results in order to estimate the risk level.

Over the last 10 years there has been a tendency to integrate management systems, such as quality and environmental management systems, or quality, environmental and occupational health and safety management systems. The most important benefits from this integration are (*a*) the reduction of documentation, (*b*) the optimization of internal and external audits, (*c*) the communication and training activities, (*d*) the improvement in efficiency of operations, (e) the saving of resources and (*f*) the reduction in bureaucracy as reported by Salmone (2008), Sanz-Calcedo et al. (2015), Santos et al. (2011) and Nunhes et al. (2017).

In addition a few attempts have been performed in order to achieve a Total Safety Management framework. This framework may either integrate occupational health and safety with operational process safety (Badri et al. 2012) or avoid risk in the short and long term by using a proactive approach and designing inherently safe processes (Zhi et al. 2012).

The methodology for Total Safety Management for industrial organizations

The Total Safety Management Framework is composed of several interrelated modules, as reported by Leva et al. (2014) as follows:

1 The Common Operational Picture
2 Risk Assessment for Design
3 Risk assessment for Operations
4 Risk Assessment for Critical Activities

The different modules comprising this framework are briefly introduced below, while more details are presented by Kontogiannis et al. (2017).

Common Operational Picture

The Common Operational Picture (COP) is a "mental model of how the system works" and guides the application of a safety management system in everyday practice. The COP has been implemented with success in Defence Departments (Liu et al. 2011), where its aim was to collect, process and manage information of real-time battlefield situation information, though it can also be applied in emergency situations that may arise in the process industry. The COP "incorporates information that enables situational knowledge to be produced, visualized and presented in a way that all information is available to all the actors involved in the crisis response in a real time", as reported by Luokkala et al. (2017). The COP could also incorporate risks that are significant in a particular area together with uncertainties that exist in the risk evaluation and risk mitigation measures of a process industry, so as to present a compound risk picture of the installation. This picture may be represented in different ways but, nevertheless, it should be accessible by all stakeholders involved in either the prevention or the mitigation phase of accidents in order to analyse and communicate risk, but also support training and procedures design. It should provide a common understanding of various types of information required and visualized for accident prevention and mitigation. The information may be gathered from a number of sources within a process plant (e.g. maintenance, operation department for equipment repairs and tests) and communicated to various decision makers and stakeholders, either internally or externally. The COP provides information related to human tasks and errors, assessed with task analysis, and overall risks assessed via quantitative risk assessment methods.

Total Safety Management Framework for design

Risk and safety management should be considered throughout the whole life cycle of a system, starting with the design phase. This view is proposed by ISO 31010 (ISO 2009), which states that risk assessment should be conducted during life cycle phases. During the design phase risk assessment contributes to ensure that system risks are tolerable and helps in the refining of the design and in the identification of risks impacting upon subsequent life-cycle phases. The consideration of safety at this stage is at the core of the Inherently Safer Design approach (Kletz 1998); Human Centred Design also seeks to create designs that minimize human errors. The Total Safety Management framework proposed integrates these approaches by supporting Inherently Safer Design with the use of new technologies such as 3D modelling simulations and at the same time supporting design review with Human Factors or Human centred approaches. Task analysis and Ergonomic reviews may improve both the physical environment (e.g. maintainability of equipment, accessibility etc.) and the cognitive aspects (e.g. human decision making, provision of information, etc.). 3D modelling simulations have been widely used for training, but also in the design of workplaces (Gardeux & Marsot 2014) in order to take into account health and safety conditions.

Total Safety Management Framework for operations

Total Safety Management Framework focuses on monitoring and management of risk during the operational phase. What is required is a clear understanding of how risks are currently monitored, for example through incident reporting or safety performance indicators, how training is used to manage risk in the operational phase, and how changes are communicated. A knowledge management system is created and various types of data are gathered from the risk assessment and the risk mitigation levels. The system functioning and operator performance data managed by the knowledge system are the following:

- Functional descriptions of sub-systems and plant equipment
- P&ID of process equipment
- Dangerous substances data
- Failure modes and effects analysis
- Operational hazards
- Existing operating procedures, checklists and work permits
- Inspection regimes and maintenance planning policies

Total Safety Management Framework for Critical Activities

Critical activities are those which may either introduce major changes to an organization or have high risk. The total safety framework seeks to understand how these activities can be managed in order to assess and reduce their risk.

The methodology and procedures followed for the risk quantification during operation and critical activities is based on the quantitative risk assessment methodology developed by Papazoglou et al. (1992) for installations handling toxic and flammable substances and can be distinguished into three major phases, which are the following: (*a*) Assessment of Plant Damage States and their Frequency of occurrence, (*b*) Consequence Assessment and (c) Risk Integration.

The "Assessment of Plant Damage States" consists of analysing the installation to identify potential accident initiators, assessing the response of the plant to these initiators and establishing end damage states of the plant resulting in the release of a dangerous substance in the environment. It can be distinguished in the following procedural tasks:

Hazard source identification

The main sources of potential hazardous substance releases that lead to loss of containment are identified and the initiating events that can cause such releases are determined. In order to identify the sources of substance release the following tasks are performed: plant familiarization and information gathering, identification of releases of concern, identification of operation phases and finally initiating event selection.

Construction of logic models

A logic model for the installation is developed in this step. The model includes each and every initiator of potential accidents and the response of the installation to these initiators. Specific Bowties that are equivalent to Event Trees are defined (Papazoglou & Ale 2007). Bowties are diagrams used to depict the logical interrelations of the various events the outcomes of which determine the accident sequences. The concept of the "bowtie" is based on the definition of an important event that describes the main characteristics of an accident. Such an event is, for example, "Fire in the dichloropropene area" (see Figure 9.1). Appropriately, this event is called the "Centre Event". Then the model can be distinguished in two parts:

- One on the left of the "Centre Event" called the Left Hand Side (LHS)
- One on the right of the "Centre Event" called the Right Hand Side (RHS).

The LHS of the model depicts the part of the model that includes the causes that led to the Centre Event. These causes are determined and modelled on a deductive way. One of the tools used to analyse causes starting from a general cause and proceeding to levels of increasing decree of detail is the "cause tree" or "fault tree". Each level of such a tree provides the logical way in which the events of this level must combine to "cause" an event of the immediately higher level. As the analysis proceeds from the composite events to simpler events or causes, the tree expands, thus forming the LHS of the "bowtie"

In principle, the LHS of a bowtie includes all the elements of the accident sequences that correspond to "Prevention" measures, that is, to whatever means exist that aim at preventing the centre event from occurring. Similarly the RHS of a bowtie includes all the elements that influence the final consequence given that the centre event has occurred and hence it models all the "Mitigation" aspects of an accident.

A fundamental concept of the bowtie model is the concept of a "safety barrier". A safety barrier is a technical, hardware, procedural or organizational element in the working environment that aims either at preventing something from happening (e.g. the CE) or at mitigating the consequences of something that has happened. Some of these events or barriers have a deterministic logical relationship with the event that follows them or the event that they cause. For example, "Outflow spill" coupled with the "presence of ignition sources" results in fire in the dichloropropene tank.

In conclusion, the bowtie model provides a way for organising various events connected either with the prevention or the mitigation of accidents into sequences of events starting from a root cause and ending up with a damage state. Pictorially, an "accident sequence" can be visualized as a path in the "bowtie" that starts from the left, goes through various events of the LHS, passes through the CE and then through various events of the RHS and ends with a final consequence at the outmost right of the bowtie (see Figure 9.1).

The use of such a model is twofold. On one hand, it provides the accident sequences, that is, the sequence of events that leads from a fundamental or root cause to the final consequence; this is equivalent to providing the "structural insight" or the deterministic relationships of various simple events combined to cause a more composite event. On the other hand, it provides a way for quantifying the risk since the assessment of the probability of the final consequence (i.e. the risk) is now transformed to the assessment of these probabilities of composite vents (accident sequences). The

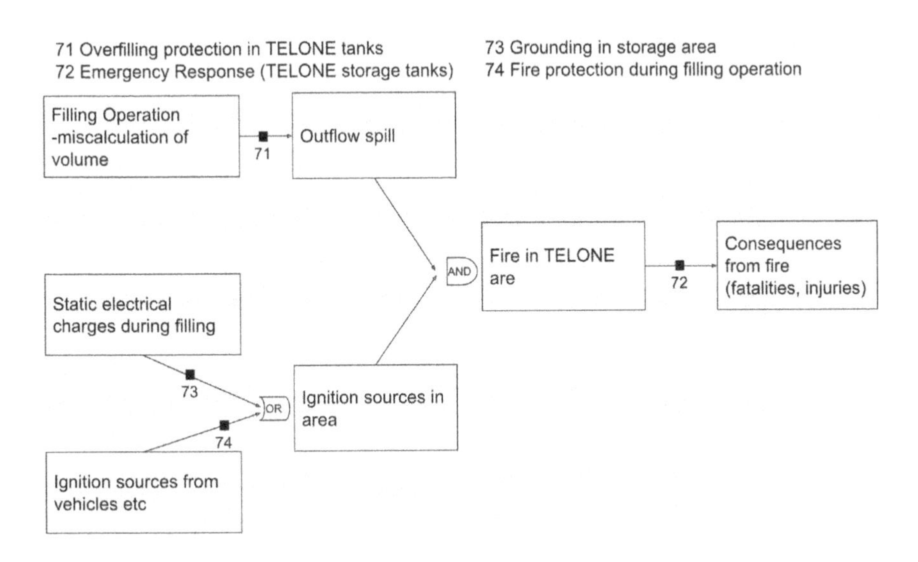

Figure 9.1 Bowtie for overfilling of storage tanks and fire in the area

latter can be now achieved following the laws of probability theory, if the probabilities of the constituent simpler events are known. Furthermore, it provides a tool through which the effect of a specific action on the risk can be calculated if the effect of this action on simple events forming the input of the model can be assessed.

The visual attractiveness of the bowtie model and its "communication power" is high. In any case the actual underlying model is an event tree. The basic character of the model is, however, the same as that of a bowtie: it provides the logical interrelationships between simpler events (technical, procedural, organizational) with the composite event of a reportable consequence of an occupational accident. These relationships are consistent with and amenable to quantification according the laws of probability theory.

Barrier or system failures can be modelled with more detailed models, namely with Fault Trees, which in terms of basic component failures and human errors allow for the quantification of the system failure probabilities and subsequent accident sequence frequencies.

Plant damage state definition

A plant damage state uniquely characterizes the installation-dependent conditions of release of the hazardous substance. Accident sequences resulting into the same conditions of release are grouped corresponding to a particular plant damage state.

Consequence Assessment and Risk Integration

"Assessment of Consequences" aims at establishing the consequences of the released hazardous substances. A set of major steps can be distinguished for the assessment of the consequences of released flammable substances, as presented by Papazoglou et al. (1992), which are the following:

Determination of Release categories, defines all necessary physical condition, phenomena and parameters that uniquely determine the level of thermal flux or the overpressure at each point in the area around the emulsion source, in case of flammable materials.

Estimation of Heat Radiation and Peak Overpressure, with models simulating heat radiation or peak overpressure, resulting from the released flammable materials.

Dose Assessment an individual receives, owing to the exposure to the extreme phenomenon generated by the flammable material.

"Consequence Assessment" assesses the probability of fatality or injury of the individual receiving the dose calculated in the previous step.

"Risk Integration" combines the frequencies of the various plant damage states with the corresponding consequences and results in the quantification of risk, by assessing individual fatality risk at the area near the dangerous substances release.

Development of a tool for Total Safety Management System

The tool developed for Total Safety Management is built around the critical equipment/tasks identification in three layers:

A First, critical equipment has been recognized during the risk analysis which consisted of the following methods: determination of critical activities, fault tree analysis, and bowties for event sequences. Next a set of attributes and requirements was assigned to each piece of critical equipment, such as the following:
- Actions related to commissioning, inspection and maintenance with a desired predefined periodicity and associated documentation (e.g. maintenance instructions)
- Related Standard Operating Procedures
- Documents of any type (texts, photos, diagrams)
- Substances involved during production
- Person(s) responsible – action owners

Equipment can be grouped in sets, either according to its location or function, giving the opportunity to review its attributes and requirements collectively.

B At the second layer, which is a 3D plant mapping environment, each equipment photo is placed in the actual plant location, helping the user to assess and manage understand it.

Furthermore, the 3D environment can be used as a virtual plant tour tool (for the time being only two locations have been implemented for this case study), which is useful as an introductory training session for new employees or contractors hired for a specific task.

Based on the Major Accident Assessment, impact zones of the various scenarios can be also depicted and presented to the various stakeholders involved (personnel, public, authorities) in a more vivid way than a hardcopy map.

C The third layer is the Computerized Barrier Management System (CBMS) tool that provides the integration of the above two layers at a user friendly Web interface:

- A user can handle all the equipment actions, as they have been set at the first level, having all the related information in one screen.
- Planning of actions is done either automatically based on the defined periodicity or manually, if a new action is inserted
- Incidents, accidents and deviations related to equipment/task can be reported using the tool, together with the appropriate forms/documents
- List of pending and planned actions is readily available based on equipment/set of equipment or person responsible, providing the ability to organize jobs more efficiently and omission free.
- History logs are organized in one screen per equipment and maintained even if the equipment is replaced with a new one (decommissioning of old and commissioning of a new)
- Each piece of equipment can be focused on the 3D environment (second layer) providing a "live" view of the place where an action is planned to take place

Given all the above, all the critical information is gathered in one place, being transparent to all and readily accessible.

The tool consists of the following modules: (*a*) The Computerized Barrier Management System (*b*) the REVIAMAPS tool (viewed November 2017, http://reviatech.com) for mapping and (*c*) the Web equipment management tool.

The computerized Barrier Management System focuses on the Common Operational Picture and Risk assessment for operations and critical activities. The general processes of the Computerized Barrier Management system are presented in Figure 9.2. It is a tool for monitoring and controlling hardware systems, managing issues related to workflow and task allocation, and it also contains a registry of all plant equipment data together with their maintenance

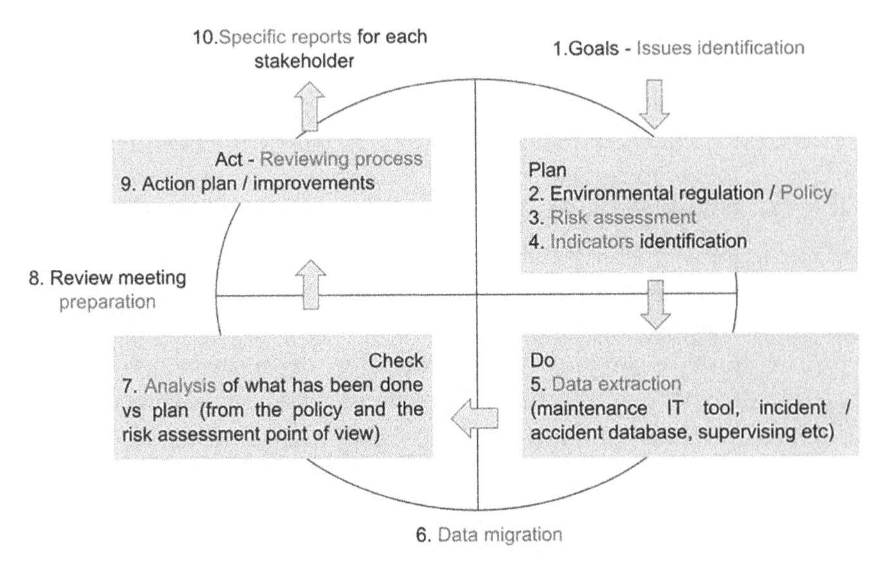

Figure 9.2 Processes of the Computerized Barrier Management System

schedule. Additionally, all risk assessment information required for a plant is gathered and displayed in a database platform. Last, but not least, it has visualization capabilities for risk representation, by drawing and showing fault tree and bowtie diagrams of a hazardous plant, which assist stakeholders to share the common risk operational picture.

The CBMS can handle the following information:

a) Accident sequences, bowties and Fault Trees in the form of graphs
b) Flowcharts of processes
c) Details of barriers, such as description and documents related to the barrier
d) Critical activity data registered, such as: organization (the personnel that carries out the critical activity), accreditation models, barriers (the barriers – tasks that are linked to the critical activity), procedures (the procedures that refer to the critical activity), restrictions/obligation models, record models, display models, proof models, equipment (the equipment that is involved in the critical activity) and graphs (graphs in which this critical activity appears)

e) lists of tasks required to perform the safe operation of barriers, list of barriers, equipment, phases of plant operation,
f) organization data of the safety management and data required for phases of plant operation or critical activities such as the following:

- Procedures and materials safety data sheets including intrinsic hazards and adequate PPE and changeover limits
- Communication between production and maintenance
- Work permits
- Risk assessments – health and safety risk assessments, environmental risk assessments, accident scenarios, consequences, etc.
- Warehouse instructions
- Incident reports
- Maintenance history and planning
- Component data, and
- Management of change (modifications)

The REVIAMAPS tool is a 3D mapping tool developed by REVIA Tech that allows graphical representation of an installation in various levels of detail. As a basic concept, the general layout of a plant is represented with buildings and rooms placed in their actual positions and sizes within the plant. Basis for the design can be an AutoCAD© or other similar tools drawing. In case of further detail, specific equipment and instruments can be also designed and represented. The user can "walk" inside the plant and look from various locations and angles of the site map, have a closer look of an item or see a picture of the item, if the latter has been added. In the paragraph presenting the Risk assessment of the tank farm containing dichloropropene several screenshots of the REVIAMAPS tool are presented.

The Web equipment management tool manages all equipment data through a dedicated Web tool. The analysts may simply log in to the tool from a PC, a tablet, or a smartphone and practically manage all actions required for equipment such as commissioning, maintenance, periodic tests, certificates and decommissioning of a specific item.

These tools may communicate, in a way that the user can simply navigate in 3D maps and by selecting a specific equipment, its relevant actions can be shown in the management tool.

Brief description of the case study: Tank farm containing dichloropropene

The case study where the Total Safety Management Framework has been demonstrated is the K&N Greek plant that produces agrochemicals, for the protection of plants, such as herbicides, insecticides, fertilizers, etc., and also plant seeds, namely for cotton, corn, cereals. The company's installations are in the industrial area of Thessaloniki, in northern Greece, with total surface 40,000 m^2 and 90 workers.

The case study consists of the tank farm, namely three tanks of dichloropropene with capacity 30 m^3 each, where drums of 20 liters of dichloropropene are produced (see Figure 9.3). In case of an accidental release,

Figure 9.3 Tank farm containing dichloropropene

separate retention bunds around the tanks may contain all the quantity of dichloropropene stored. Tanks are equipped with pressure relief valves, pressure protection and a firefighting system. The tanks are loaded with dichloropropene through special containers called "isotanks", which are transferred to the installation by trucks. There are three operating phases of the storage tanks, namely loading, storage and unloading.

Risk assessment of the tank farm containing dichloropropene

The major phases of the risk assessment methodology, presented in paragraph "Total Safety Management Framework for Critical Activities" are applied in the case study as follows.

Hazard source identification

A HAZOP and a detailed Master Logic Diagram method, as described by Papazoglou and Aneziris (2003), were applied to the case study storage tanks during all operational phases of the tanks (storage, loading, unloading). Both methods, when applied so as to identify direct causes leading to Loss of Containment (LOC) in the dichloropropene storage tanks during storage obtained the following events: (*a*) external fire and (*b*) earthquake. Initiating events, or direct causes, leading to LOC during the loading the tanks are the following: (*a*) high level in the tanks, (*b*) external fire and (*c*) containment bypass. Finally initiating events (IE) identified for the unloading phase of the tanks are: (*a*) external fire, (*b*) low level of the tanks and (*c*) leak from the unloading pipe. These initiating events may lead to the five damage states presented in Table 9.1.

Once the initial list of IEs has been compiled the next methodological step consists in the determination of the safety functions and the systems that serve these functions. Safety functions and systems are incorporated in the design of the facility to prevent and/or mitigate the possible consequences of the IEs. The systems that serve the safety functions directly are called frontline systems. These are the systems that will form the headings of the event trees developed later. Safety functions and systems are presented in this section for the case study.

Table 9.2 presents the safety functions incorporated in the design of the plant. Systems that have as mission to perform the identified safety functions

Table 9.1 List of initiating events and plant damage states of the case study

Initiating event	Plant damage state
Miscalculation during filling of tank	Outflow spill of in storage area
External Fire in storage tanks during storage – unloading	Tank rupture (overpressure)
External Fire in storage tanks during filling	Tank rupture (overpressure)
Earthquake	Tank rupture
Containment by pass	Spill in storage area
Low level in tank	Tank rupture (implosion)
Leak from pipe	Release from pipe
External Fire near isotank	Rupture of isotank

Table 9.2 List of Safety Functions

Isopropene storage tanks
Avoid overfilling of storage tanks
Provide fire protection during filling
Provide grounding protection of storage tanks
Avoid overpressure of storage tanks
Provide stop of loading of tanks in case of containment bypass or leakage
Provide stop of unloading of tanks in case of low level
Avoid low level in dichloropropene tank
Provide emergency response

are given in Table 9.3 for the dichloropropene tanks. Table 9.3 provides in summary form the corresponding success criteria for each frontline system. Success criteria are the minimum performance requirements for a frontline system to successfully provide the associated safety function. As an example, we mention the pressure safety valves of the dichloropropene tanks. They serve the safety function of avoiding overpressure. The success criteria for serving the former safety function is successful operation of one valve.

Accident sequence determination

The completion of the Hazard Identification methodological step resulted in the identification of eight initiating events, which require the determination and assessment of eight bowties, presented in Table 9.4. Figure 9.1 presents

Table 9.3 Safety Systems and success criteria of frontline systems

Frontline Systems	Success criteria
High level tank control system	Level sensors, indicators and alarms, operator
PSV of dichloropropene tanks	One out of two
Grounding system of storage tanks	
Firefighting system of storage tanks	Water tank, pipe, pump, sprinklers
Emergency stop of loading	Communication between plant operator and isotank operator
Emergency stop of unloading	Operator in plant
Low level control system of dichloropropene tanks	Level sensors, indicators and alarms, operator
VSV of dichloropropene tanks	One out of two
PSV of isotank	One out of two

Table 9.4 List of bowties developed for risk assessment of the storage area

Bowties for risk assessment of storage area
Overfilling of dichloropropene tanks
External Fire in dichloropropene tanks during storage/unloading
External Fire in dichloropropene tanks during filling operation
Earthquake
Containment bypass before loading of dichloropropene tanks
Low level in tank while unloading tanks
Leak from pipe in the dichloropropene area
External fire near Isotanks

a bowtie with initiating event high level in the tank, caused by miscalculation of the volume of the dichloropropene, which is loaded in the tanks. The barrier which prevents the tank from overfilling is the overfilling protection system. In case this barrier fails, an outflow spill will occur and if an ignition source is encountered, a fire will occur in the storage area. Ignition sources can be avoided, if one of the following barriers operates: grounding of the tanks or fire protection. In case of a fire in the storage area, the only barrier to prevent a major catastrophe is the effective application of the pre-existing "emergency response plan of the installation".

The overfilling protection system has been analysed into simpler events, in order to assess the failure rate of the former with the Fault Tree methodology. Figure 9.4 presents the Fault Tree that has been developed for overfilling protection; it can be noticed that overfilling protection fails only if the operator does not stop loading when required. This may occur either if the operator doesn't receive a high level indication or a high level alarm, or if the pump loading the tank fails to stop, or because of a human error the operator ignores the high level signals. If the level sensor has failed the operator will not receive a high level signal, while if both level alarm and level indicator have failed he will not get any signal. In this tool there is the additional capability to assign "tasks" to each basic event. The completion of these tasks is required for the safe operation or maintenance of the relevant equipment. For example, maintenance of the level alarm is a task which should be performed and scheduled in order to have a reliable and safe piece of equipment, as presented in Figure 9.4.

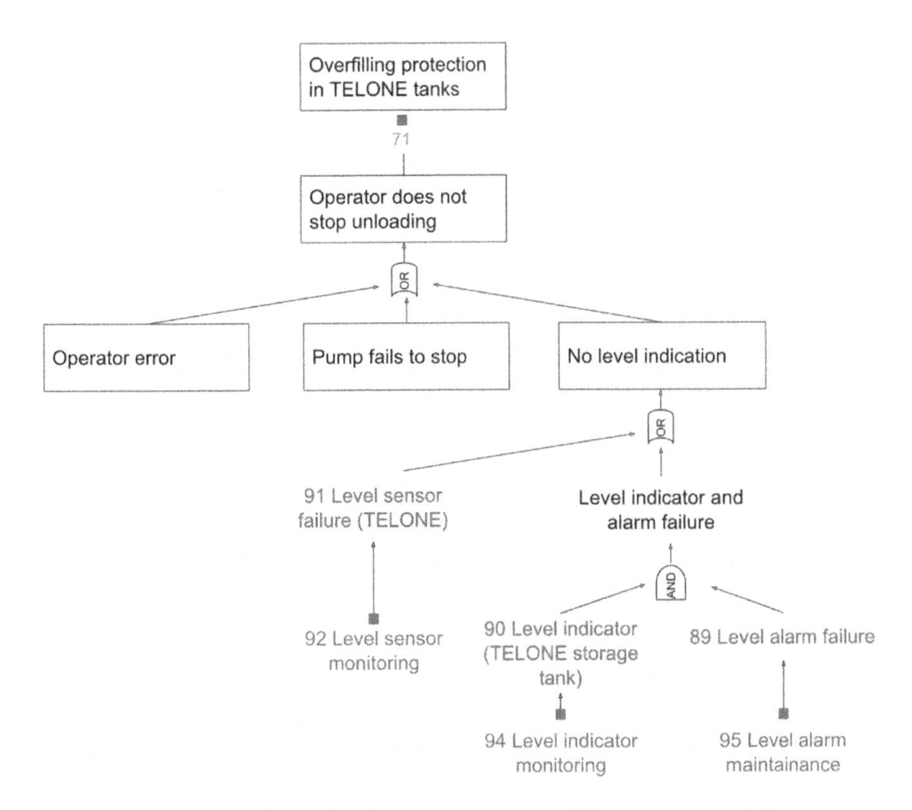

Figure 9.4 Fault Tree for the event of the overfilling protection failure

Figure 9.5 Screen with inside view of rooms and related equipment in REVIAMAPS

Table 9.5 List of "Top Events" of Fault Trees developed for risk assessment of the storage area

"Top Events" of Fault Trees developed for risk assessment of storage area
Failure of overfilling protection system
Failure of Emergency response
Failure of operators to stop loading
Failure of overpressure protection of dichloropropene tanks
Failure of low level control in storage tanks

The tasks which can be assigned to safety equipment according to the I-RISK methodology presented by Papazoglou et al. (2003), are the following: provide, use/operate, maintain and monitor/supervise safety barriers. In total five Fault Trees have been developed and quantified for risk assessment of the storage area, as presented in Table 9.5.

Consequence Assessment

Heat radiation in case of fire in the area of the storage of flammable dichloropropene has been assessed with a tank fire model simulating heat radiation (Papazoglou et al. 1996). For the first four damage states presented in Table 9.1, namely, outflow spill of dichloropropene in storage area or dichloropropene tank rupture either owing to overpressure or to implosion and release from pipe, it has been assumed that 35 t of the flammable substance will be released and immediately ignited in the bund area of 65 m^2. In this case the level of conditional individual risk will be equal to 10^{-2} in an area of 80 m^2 around the release, or at a distance of 5 m from the point of release, given than an individual remains for 60 s in that spot. In case of the "isotank" rupture, conditional individual risk has the same values, if the radius of the pool fire is assumed to be the same as in the previous case.

Risk Integration and visualization

In order to visualize risk a 3D plant mapping environment has been developed and each equipment photo has been placed in the actual plant location, helping the user to assess, manage and understand it (see Figures 9.5, 9.6 and 9.7). Based on the Consequence Assessment, impact zones of the various scenarios are also depicted and presented to the various stakeholders involved (personnel, public, authorities) in a more vivid way than a hard-copy map, as presented in Figure 9.7. Furthermore, the 3D environment can also be used as a virtual plant tour tool.

Information handled in the database for barrier management and handling safety

The elements required for risk assessment and safety management of the case study are the following: (*a*) Material Safety Data Sheets with handling instructions for dichloropropene, (*b*) documents of any type such as texts, photos, PID diagrams for the storage tank, (*c*) procedures for reception, handling and storage of incoming material, equipment maintenance, personnel training, risk assessment and unloading, (*d*) restrictions and obligations in certain areas, such as no smoking, no traffic and use of personnel protective equipment and (*e*) safety equipment such as the firefighting system in the storage area (foam tank, sprinklers and gas detectors) and all other equipment required for the operation of the storage tanks. This information can also be viewed for each operation phase, as for example for the storage phase of the tanks, presented in Figure 9.8. Specific elements which have to be managed are the tank sprinklers, the grounding of the tanks and the

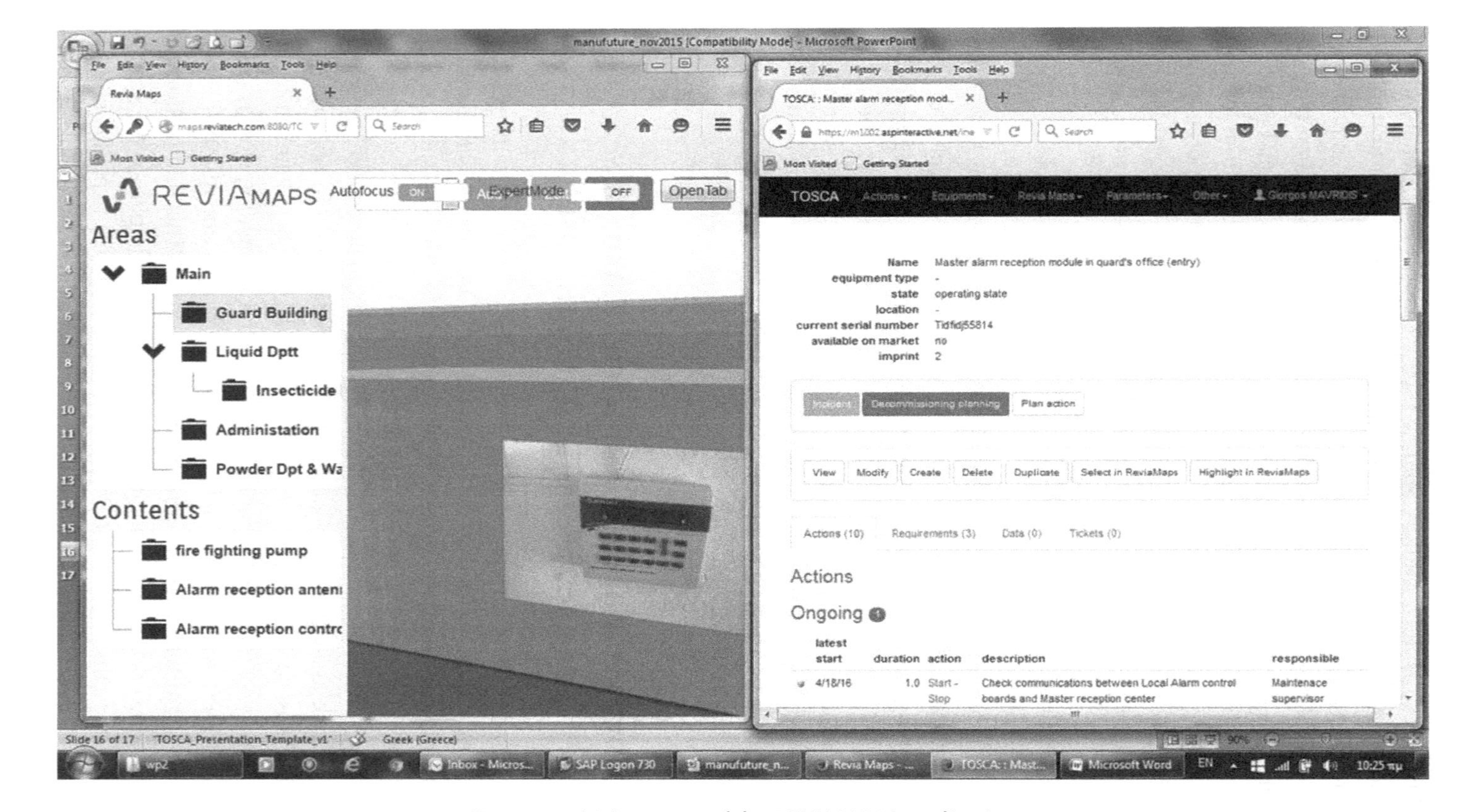

Figure 9.6 Main screen of the REVIAMAPS application

Figure 9.7 Layout of industrial plant

pressure valves. Relevant procedures and restrictions in the storage area are also presented in Figure 9.8. In addition, a list with all equipment of the storage facility may be provided, while information regarding actions related to commissioning, inspection and maintenance together with responsible operators is also available.

All data related to the equipment that was mentioned in the previous paragraphs are also managed through a dedicated Web tool specifically developed in TOSCA project. The user should simply log in to the tool in any device (PC, tablet, smartphone) and practically manage all actions required for his equipment (commissioning, maintenance, periodic tests, certificates, decommissioning) as well as related incidents.

The notion of functionality of this tool is that the user can add equipment of interest, usually but not only involved in accident sequences, critical activities and a barrier level of confidence, as they have been detected and inserted in the CBMS database. The data that are displayed on the web tool refer to:

- Equipment data, as name, location, state, serial number
- Actions performed to equipment
- Requirements (e.g. planned maintenance or inspections)
- Data (e.g. photos or associated documents)

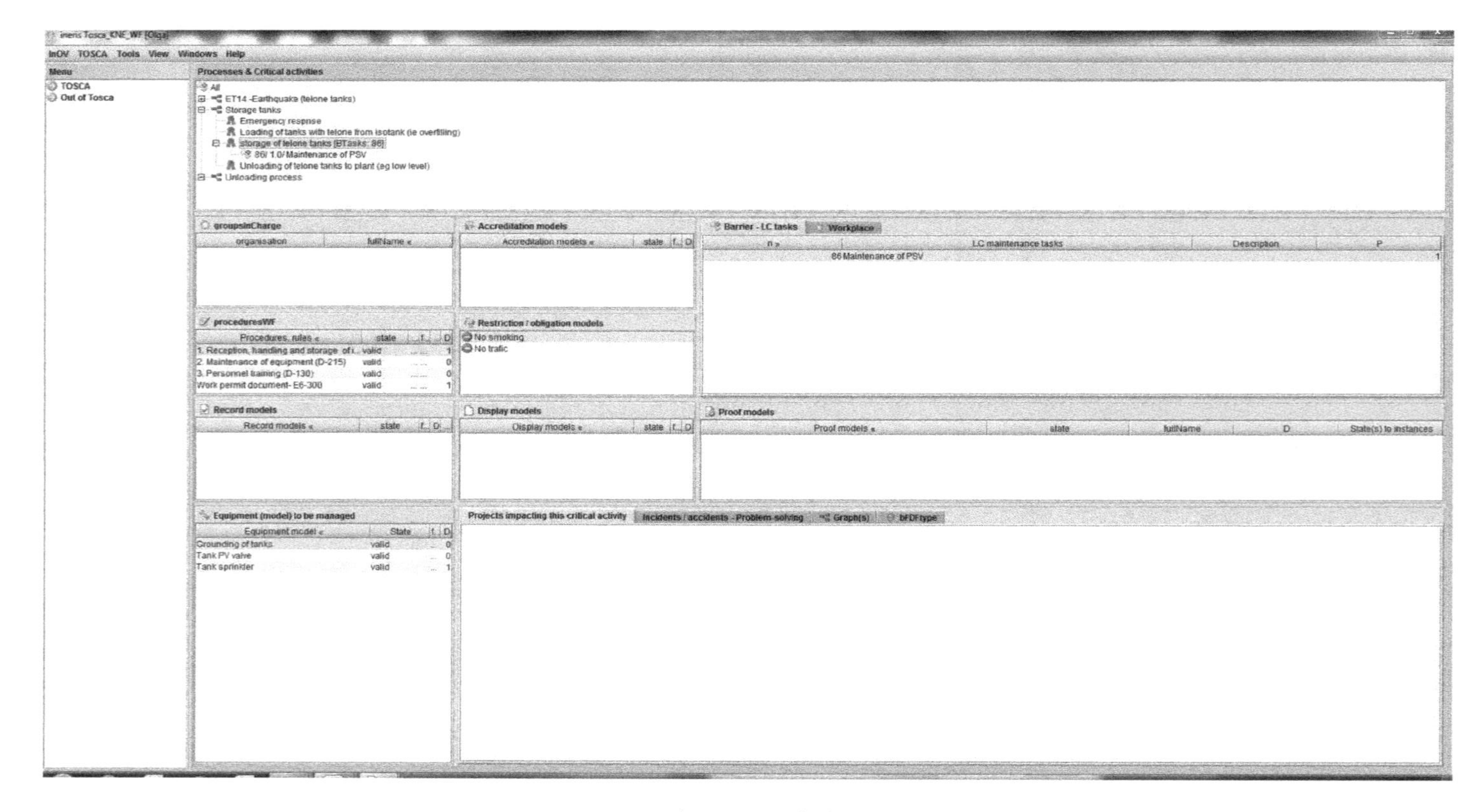

Figure 9.8 Management of operational phases or critical activities

The actions shown on the Web tool are divided in three sections:

- Ongoing and planned actions: actions on equipment that have been planned for the future or actions in process (but not yet overdue)
- Pending actions: overdue actions assigned to the equipment
- History: all actions assigned to the equipment

Discussion of results

The aim of this tool was to provide a Common Operational Picture for all stakeholders involved in the daily plant operation, in a real-time manner that keeps all relevant data integrated, recorded, transparent and easily accessible. Before the introduction of the Total Safety Management System tool, the information in the K&N plant was dispersed in various pools and forms such as electronic in the computer server or as hard copies. This arrangement was difficult to access, cross-check and update in a coherent way. Modifications or interventions might occur, without taking into consideration all aspects related to certain equipment and its criticality to overall safety. Furthermore planning of inspections and maintenance was performed manually and a consistent and integrated log file containing all actions and relevant data was not available. All the above actions were time consuming and prone to errors and omissions. The e-solution implemented in this case study was evaluated against the aim to tackle the above issues in a more efficient way.

The application may be expanded to an entire plant. Assistance is required for the 3D plant representation of the plant buildings. This can be either on a higher level of abstraction (without detailed drawings of each room of a building) or more detailed (based on the actual layout and schematics of each building). Risk analysis and equipment attributes and requirements will have to be expanded to the whole plant.

Applicability and usability to SMEs

The major parts of the methodology developed for Total Operations Management for Safety Critical Activities has been presented in this chapter can be readily applicable and useful for other SMEs of the greater chemical and petrochemical domain, the food industry and the pharmaceutical industry.

The benefits of the methodology and associated tool are that all risk and safety related information can be incorporated into a database, with graphical interface capabilities. The aim of this methodology and associated tool is to facilitate the everyday plant management including, standard operation procedures, preventive inspections and preventive and corrective maintenance actions. The whole process of plant (equipment) state and emergency management are transformed to an e-solution tool, while the integral IT tool communicates also with external visualization tools.

In addition this innovative tool has been also transformed in a user friendly Web interface which may provide help, especially for SMEs in the following aspects: (*a*) include parts of the management system, necessary for the plant safety, (*b*) include the risk analysis for various plant phases and critical tasks, (*c*) support the risk model implementation with a planning management system concerning even daily activities, (*d*) help the management of actions/decisions/meetings, (*e*) allow cross-checking between actions to do, actions done, pending issues, and incidents or accidents, with the easiness of a dashboard, (*f*) permit safety plant monitoring, (*g*) facilitate planning of equipment actions and providing with the history logs concerning maintenance, inspection, commissioning etc., (*h*) provide a "live" view of equipment and the place where an action is planned to take place and (*i*) offer access to information and reporting on incidents accidents and deviations.

In summary, the specific framework and associated application aims at developing a risk improvement tool and a knowledge management tool that can be used for integrating risk management, maintenance activities and daily operations. The risk analysis tool aims at identifying hazards and human barriers that are important in the context of maintenance and operations. The knowledge management tool assists analysts in accessing and managing all risk information in the organization. It can also provide a basis for workplace prototyping so as to allow the creation of an integrated knowledge base for work best practices and safety requirements to be shared across the organizational units.

Acknowledgements

The work presented in this paper was performed in the scope of EU 7FP project TOSCA under grant agreement FP7-NMP-2012-SMALL-6–310201.

References

Ale, B, van Gulijk, C, Hanea, A, Hanea, D, Hudson, P, Lin, P-H & Sillem S 2014, 'Towards BBN based risk modelling of process plants', *Safety Science*, vol. 69, pp. 48–56.

Aneziris, ON, Papazoglou, IA, Konstandinidou, M & Nivolianitou, Z 2014, 'Integrated risk assessment for LNG terminals', *Journal of Loss Prevention in Process Industries*, vol. 28, no. 1, pp. 23–35.

Badri, A, Gbodossou, A & Nadeau, S 2012, 'Occupational health and safety risks: Towards the integration into project management', *Safety Science*, vol. 50, pp. 190–8.

CCPS–Center for Chemical Process Safety 2000, *Guidelines for chemical process quantitative risk analysis*, 2nd edn, American Institute of Chemical Engineers (AIChE), New York.

Gardeux, F & Marsot, J 2014, 'A 3-D interactive software tool to help VSEs/SMEs integrate risk prevention in workplace design projects', *Safety Science*, vol. 62, pp. 214–20.

ISO 2009, *ISO 31010–risk management–risk assessment techniques*, International Standards Organisation, Geneva.

Khan, F & Abbasi, SA 1998, 'Techniques and methodologies for risk analysis in chemical process industries', *J. Loss Prev. Process Ind.*, vol. 11, pp. 261–77.

Kletz, TA 1998, *Process plants: A handbook for inherently safer design*, Taylor and Francis, Philadelphia, PA.

Kontogiannis, T, Leva, MC & Balfe, N 2017, 'Total safety management: Principles, processes and methods', *Safety Science*, In Press, Available online 4 October 2016.

Leva, MC, Balfe, N, Kontogiannis, T, Plot, E & De Michela, M 2014, 'Total safety management: What are the main areas of concern in the integration of best available methods and tools', *Chemical Engineering Transactions*, vol. 36, pp. 559–64.

Liu, B, Wu, CH & Liu, HJ 2011, 'Research on the human factors of common operational picture', *Procedia Engineering*, vol. 24, pp. 182–286.

Luokkala, P, Nikander, J, Korpi, J, Virrantaus, K & Torkki, P 2017, 'Developing a concept of a context-aware common operational picture', *Safety Science*, vol. 93, pp. 277–95.

NORSOK Z-013 2010, *Risk and emergency preparedness analysis*, 3rd edn, Norwegian Technology Center, Oslo, Norway, www.standard.no/en/sectors/energi-og-klima/Petroleum/NORSOK-Standard-Categories/Z-Risk-analyses/Z-0132/

Nunhes, TV, Barbosa, LCFM & Oliveira, OT 2017, 'Identification and analysis of the elements and functions integrable in integrated management systems', *Journal of Cleaner Production*, vol. 142, no. 4, pp. 3225–35.

Papazoglou, IA & Ale, BJM 2007, 'A logical model for quantification of occupational risk', *Reliability Engineering & System Safety*, vol. 92, no. 6, pp. 785–803.

Papazoglou, IA & Aneziris, ON 2003, 'Master logic diagram: Method for hazard and initiating event identification in process plants', *Journal of Hazardous Materials*, vol. A97, pp. 11–30.

Papazoglou, IA, Aneziris, O, Bonanos, G & Christou, M 1996, 'SOCRATES: A computerized toolkit for quantification of the risk from accidental releases of toxic and/or flammable substances, integrated regional health and environmental risk assessment and safety management', *International Journal of Environment and Pollution*, vol. 6, nos. 4–6, pp. 500–33.

Papazoglou, IA, Bellamy, LJ, Hale, AR, Aneziris, ON, Ale, BJM, Post, JG & Oh, JIH 2003, 'I-Risk: Development of an integrated technical and management risk methodology for chemical installations', *Journal of Loss Prevention in the Process Industries*, vol. 16, no. 6, pp. 575–91.

Papazoglou, IA, Nivolianitou, Z, Aneziris, O & Christou, M 1992, 'Probabilistic safety analysis in chemical installations', *Journal of Loss Prevention in Process Industries*, vol. 5, no. 3, pp. 181–91.

REVIAMAPS, viewed November 2017, http://reviatech.com

Salmone, R 2008, 'Integrated management systems: experiences in Italian organisations', *Journal of Cleaner Production*, vol. 16, pp. 1786–806.

Santos, G, Mendes, F & Barbosa, J 2011, 'Certification and integration of management systems: The experience of Portuguese small and medium enterprises', *Journal of Cleaner Production*, vol. 19, pp. 17–18, 1965–74.

Sanz-Calcedo, JG, González, AG, López, O, Salgado, DR, Cambero, I & Herrera, JM 2015, 'Analysis on integrated management of the quality, environment and safety on the industrial projects', *Procedia Engineering*, vol. 132, pp. 140–5.

Tixier, J, Dusserre, G, Salvi, O & Gaston, D 2002, 'Review of 62 risk analysis methodologies of industrial plants', *Journal of Loss Prevention in the Process Industries*, vol. 15, pp. 291–303.

TOSCA website, www.toscaproject.eu/

Villa, V, Paltrinieri, N, Khan, F & Cozzani, V 2016, 'Towards dynamic risk analysis: A review of the risk assessment approach and its limitations in the chemical process industry', *Safety Science*, vol. 89, pp. 77–93.

Zhi, X, Lai, C & Zhang, C 2012, 'Model of total safety management (TSM) and its application', *Procedia Engineering*, vol. 45, pp. 204–7.

Effective virtual reality training for safety critical activities in the process industry

Tom Kontogiannis, Marko Gerbec and Mehdi Sbaouni

Introduction

This chapter explores the potential use of new information technologies in safety previews and in process control training. Small and Medium sized Enterprises (SMEs) operate under many financial and resource constraints and this is reflected in the restricted use of new technologies in safety training. Virtual reality has been a promising technology for representing plant sites and technical equipment in ways that resemble reality as well as for training process control skills in a safe manner. For instance, operators can master safety critical tasks, with or without supervision, by manipulating 3D equipment, without threatening the integrity of the plant. At the same time, operators can practice their tasks many times until mastery without the additional psychological stress of inflicting equipment damage. This absence of stress in virtual reality enhances learning and training effectiveness.

VR systems can integrate different technologies that allow users to interact with the virtual world in a multi-sensory manner. Some multi-modal VR systems allow physical interaction with virtual objects using a haptic device integrated with the computer's vision system. A haptic device is an electromechanical device that provides tactile feedback by applying forces in accordance with the user's hand motions. In an earlier chapter, VR systems have been used for developing a Computerized Barrier Management System in a major hazards company. Risk assessment requires a large number of documents, procedures and equipment drawings that are hard for non-specialists to understand and engage with. Desktop VR systems create a more accessible world and help engage staff whose views are otherwise difficult to collect.

This chapter examines the use of VR systems in industrial training as it offers a range of benefits compared to traditional training. Virtual reality allows learning-by-doing when operational constraints prevent the use of the real environment (e.g., machine availability, safety, time or cost constraints). Furthermore, VR systems can provide extra cues, not available in the real world, that can facilitate learning (e.g., visual, auditory or haptic cues); and they allow simulating the task in a flexible way to adapt it to users' needs and training goals. VR systems have been used to provide practical training in a number of human activities, including medical training (Aggarwal et al. 2006; Cosman et al. 2002; Seymour et al. 2002), manufacturing (Lin et al. 2002), maintenance (Gutierrez et al. 2012), chemical process control (Colombo & Golzio, 2016) and firefighting (Tate et al. 1997).

An important thesis of the TSM approach has been that safety methods and tools should not be developed in isolation because this could increase the cost of use and the cost of transfer of data. In order to develop a VR training tool that would be integrated with other risk assessment tools, a design framework is needed that examines how to screen tasks for training, how to develop the VR environment, how to deliver training and how to assess the mastery of skills. In previous chapters, it has been shown how methods of task analysis and risk assessment can be supported with the use of VR systems – e.g., preview rare procedures, identify hazards in virtual workstations and link safety barriers to 3D maps of plant sites. In this way, many parts of the risk assessment process are visualized and become accessible to non-specialists who need to contribute some knowledge to this process. It is reasonable then to exploit further the methodologies of other TSM pillars in order to develop training programs for operators that would support the acquisition and transfer of process control skills.

This chapter is structured as follows. The first part provides a general introduction to the use of virtual reality in industrial safety and operator training. Following this, a short review is made of studies that provided some evidence that the results of VR training can transfer to the real world and support human performance in complex environments. This leads to the main body of this chapter that presents a design framework for developing and delivering VR training in a way that maximizes the use of risk assessment methods from other TSM pillars. Finally, a case study is presented from a Slovenian major hazards site that has been used to design VR training for a loading scenario of a car tanker.

Virtual reality systems for safety previews and industrial training

Virtual reality (VR) is a representation technology that involves real-time simulation and user interaction through many sensory channels such as visual, auditory, tactile and smell. A VR system comprises a computer processing unit, a 3D representation of a real environment and a number of peripherals devices (e.g., eye goggles and other interaction devices). As VR systems are strong in spatial and visual representation of real equipment, the tasks which are most likely to benefit from training are those relying on visual perception and procedure execution such as navigational tasks, inspection tasks and procedural tasks. Unlike other training media, VR systems offer users many opportunities for experiencing 'presence' and 'immersion', the feelings of being inside in the virtual world.

VR systems may use different technologies for system representation and interaction which can be classified into four categories: desktop, projected, semi-immersive and immersive. In desktop VR systems, the real system is displayed on a graphical monitor, where special glasses can be worn for stereoscopic view, a 3D mouse for controlling navigation and gloves for manipulating objects. In projected systems, several displays may be used with stereo images allowing users to walk freely inside the visual world with the support of stereo glasses. A head tracking system continuously adjusts the stereo projection to the current position of the viewer. In semi-immersive systems, a combination of a physical model and a virtual world is often employed. For example, an operator may use a helmet-mounted display to view the artificial world and a physical device to control the system. Finally, immersive virtual reality allows users a stereoscopic view of the system (e.g., with a helmet-mounted display) as well as free movement and object manipulation in the visual system. In general, immersion provides the most natural way to visualize and navigate through a 3D world as the users only have to turn their heads in the direction they want to view, which is more intuitive than using a joystick.

According to Balfe et al. (2014), VR systems can provide valuable tools for improving industrial safety by the following means:

- *Optimizing equipment design*. Virtual reality creates realistic environments in which to test how equipment responds to variations in their use (e.g., weather conditions, mechanical failures, operator violations).

By performing many simulations, each with different variables, an understanding can be developed of the impact of such variables on overall system performance.

- *Previewing safety critical activities*. In many cases, operators may be unfamiliar with certain tasks because they are practiced infrequently (e.g., start-up or shut-down procedures). Virtual workstations can be designed with desktop VR systems for testing complex procedures by allowing operators to manipulate process controls, have a stereoscopic view of equipment and navigate on 3D images of plant sites. In this way, operating procedures can be tested before applied in the real world, hence avoiding delays and errors in execution. Of course, VR systems can also be used for providing opportunities for operators to identify hazards at their workplace.
- *Training process operators*. VR systems provide realistic representations of plant sites and working conditions which can provide a basis of industrial training. Apart from visualizing process equipment, VR systems allow multi-modal interaction; for example, operators can hear noise from the machines, feel hot surfaces and sense vibrations from the floor or the machines. In this respect, training can be based on a virtual environment to provide practice opportunities to operators. Trainees can receive various forms of instruction such as procedural guidance, verbal support by supervisors and memos popped up above unfamiliar equipment.
- *Enhancing safety management*. VR systems can be used for developing Knowledge Management Systems that support risk assessment (see Chapter 9). For example, information about the maintenance status, risks and procedures related to individual pieces of equipment can all be linked to the virtual objects in a 3D map, allowing operators to access the information by navigating through the 3D map of the plant.

This chapter focuses on VR applications to industrial training as it offers some advantages over other media-based training. VR systems enable people to make 3D representations of the real world which allows users to perceive and interact with the system in ways that are not possible with other media. For example, videos and computer animation may provide realistic portrayals of sequences of tasks and animations of procedures; however, these sequences and images are created offline and restrict real interaction with the system. By contrast, VR systems allow users to enter into the world, perceive the surroundings from different perspectives while moving through the environment or manipulating objects in the world. This type of interaction allows a more faithful representation of the real world.

In general, immersive VR offers a more faithful representation of the real world and provides more training opportunities than the other types of virtual reality. However, this comes at a high cost in terms of financial investments in computer equipment and human resources for system development. At the opposite side, desktop VR systems are easier to set up and transport than other semi or complete immersive systems. They can also project the system onto a large wide-screen display for presentation with several trainees and they can be used in the World Wide Web, which makes the system accessible by multiple users. Desktop VR, however, does not use the full potential of 3D capabilities which restricts the types of skills that can be learned. For instance, desktop VR systems do not allow users to learn how to master manual dexterity skills that are required in many maintenance and process control tasks.

Transfer of skills from virtual to real worlds

Although VR systems seem to offer a promising opportunity for industrial training, surprisingly, only a few studies have empirically evaluated the efficiency and effectiveness of such platforms compared to traditional methods. This section presents a short review of studies that demonstrate a beneficial effect of VR-based training over other traditional forms in a variety of domains including, manufacturing, maintenance, chemical process industries, firefighting and medicine.

One of the first studies that provided some evidence of the superiority of VR training over other forms of traditional training comes from the firefighting domain. In particular, Tate et al. (1997) have reported a study on how VR systems can improve firefighting performance on a decommissioned Landing Ship Dock (USS *Shadwell*) which had been used for full-scale fire research investigations into damage control and ship survivability. Twelve enlisted naval firefighters took part in the *Shadwell* study, split into two groups – 'traditional training' and 'VR based training'. Following conventional task briefings, the traditional group performed their allotted tasks onboard the *Shadwell*, and performance measurements (e.g., time, way-finding errors, etc.) were taken. The VR group used immersive VR technology to familiarize themselves with the layout of the ship and to rehearse their assigned tasks, with and without simulated smoke and fire. They then performed their tasks aboard the *Shadwell*, and the same performance measurements were taken. The firefighting task involved the following activities: locate breathing apparatus, assemble and direct the fire-attack team, find and prepare the designated fire hose, and locate and extinguish the fire. Results showed that there was an improvement in the performance of firefighters that used VR training over firefighters without VR training, in most aspects of the firefighting scenario.

Other studies on the use of VR training for maintenance skills provided additional evidence for the benefits of VR systems. Sebok et al. (2002) compared two VR-based training systems for industrial maintenance, guided and exploratory, and conventional map-based training; they found some advantages for exploratory VR training compared to conventional training. Another study with a VR platform using maintenance tasks by Gutierrez et al. (2012) found that VR training enabled novice technicians to recover more errors than the technicians who received traditional training. Similar results were found in the area of manufacturing tasks and the use of lathes and other machines. Lin et al. (2002) developed a Virtual Reality based Training System (VRTS) to train technicians how to use a CNC milling machine. An experiment compared a group of subject trained with the VRTS with a group of subjects who read an operating manual before their transfer to the operation of the real CNC milling machine. Training with VR resulted in fewer errors in the operation of the actual equipment than simply reading the operating manual. Another human factors–oriented analysis of VR training systems (Nathanael et al. 2016) found that students who received training in a virtual CNC machine outperformed another group who received traditional training when transferring to the operation of the real CNC machine.

Probably the greatest evidence for the beneficial role of VR in training comes from a study by Colombo and Golzio (2016) where a number of students in chemical engineering were trying to control a pool fire in the separation unit of an oil refinery. One group of students was trained with a static visual presentation of the separation unit whilst the second group received training in immersive reality. VR training outperformed traditional training on a large number of performance measures including number of help hints requested by trainees, speed of leak identification, reporting of problem to supervisors, and speed of controlling the consequences.

The greatest support for the beneficial role of VR systems in training comes from the domain of medicine. Strong evidence exists to demonstrate that learning of surgical skills may be achieved in virtual environments (Aggarwal et al. 2006). In laparoscopic surgery, for example, Seymour et al. (2002) and Cosman et al. (2002) described studies in which laparoscopic surgical trainees who received VR simulator training were found to have superior performance during their first real-world performance of a laparoscopic task. In the context of orthopaedic surgery, Vankipuram et al. (2010) presented a virtual orthopaedic drilling simulator that was designed to provide visio-haptic interaction with virtual bones. Through the multi-tiered testing strategy it was shown that the VR simulator was able to produce a learning effect that transfers to real-world drilling.

In general, there appears to exist a widespread belief that the main challenges for VR training effectiveness have to do with the necessary physical fidelity to mimic the resolution of the physical world. However, many studies that have relied solely on the physical fidelity of VR systems did not manage to improve training effectiveness. Technical refinements and advancements will always be welcome, but they alone cannot guarantee successful training. Gopher (2012) proposes that the value of a training system should be judged (1) by its ability to provide relevant experience, (2) by the provision of facilitation and guidance to the acquisition of the designated skill, and (3) by the transfer from VR training to performance in the real world. In fact, lack of human factors input has been a common cause of failure in the development of virtual reality systems. To become effective, then, VR training should be based on a thorough task analysis of visual cues of the environment, hazards at the workplaces, feedback information about work progress, opportunities for error recovery, procedural skills and decision-making skills. For this reason, the methodologies of task analysis and risk analysis should be used as a guiding framework for designing key features of virtual reality, for delivering training courses and evaluating the effectiveness of VR-based training.

A design framework for delivering VR-based training

This section describes a case study in a Slovenian major hazards plant (i.e., Plinarna Maribor d.o.o.) that demonstrates a risk-based framework for designing and delivering operator training based on virtual reality. Although industrial training belongs to the fourth TSM pillar, the methodologies that have been used for the hazard identification and the risk management pillars can still provide a framework for exploiting operator training. In this respect, all TSM pillars are organized around the same methodologies in order to optimize the technical and human resources employed in a case study.

Case study description

This case study has looked into a common procedure of loading or unloading a car tanker with a dangerous material (i.e., LPG) as a test-bed for testing the capabilities of virtual reality in industrial training. The first stage involved the definition of the objectives of the case study so that it responds to the needs of the TOSCA project and the needs of the particular plant. As a result of a consultation with the industrial users, the case study covered not only the objective of operator training, but also the need to use the same system for performing safety previews of tasks within a team environment to visualize hazards before the job is done.

The overall management of this case study was assigned to Technical University of Crete (TUC) that was responsible for defining the operational tasks for training, the scope of virtual reality and the modes of VR-based training. The Jozef Stefan Institute (JSI) acted as a technical advisor for the task analysis and risk assessment, while INERIS provided software capabilities for carrying out the risk assessment. Finally, Reviatech developed the VR prototype and the Plinarna Maribor plant delivered and assessed training in process control operations. Furthermore, the end users at the Plinarna Maribor site requested that research teams demonstrate how the training system could be integrated within the TSM framework and how to reduce the costs of the training system.

The plant site conditions involved the manual operation of LPG loading in a car tanker, a task that usually required a driver and a site operator. The operator was responsible for the manual alignment of valves in two buildings while the driver was responsible for connecting the car tanker to the installation and for the overall operation. However, it was agreed that a simplified version would be produced involving only one person (e.g., the car tanker driver) responsible for the loading scenario whilst the tasks of the site operator could be automated in the VR environment by allowing the same person to perform valve alignments at distant locations on a virtual panel that appeared on the VR world. This simplification has speeded up the whole process, hence avoiding unnecessary delays that could be introduced by the long travel distances to adjust valves from the LPG storage tanks.

VR design framework for training

As argued earlier, industrial training should not be developed in isolation from other TSM pillars because this would increase the cost of training and the cost of transfer of data between methods used in different TSM pillars. Hence, methods for VR training should be integrated with other risk assessment methods. For this purpose, a design framework is presented below that examines how to screen tasks for training, how to develop the VR environment, how to deliver training and how to assess the mastery of skills. Figure 10.1 shows the main stages of the VR design framework as summarized below:

1 Perform risk assessment to identify deviations from safety procedures. For this purpose, task analysis can be performed to describe human interactions with the technical system and provide valuable input to risk assessment. Task analysis can be carried out on spreadsheets, flowcharts, hierarchical documents or dedicated software (e.g., the SCOPE tool in Leva et al. 2015). In addition, task analysis may require data related to the design and use of process equipment that can be obtained from plant documents, design documents and Pipe & Instrumentation Diagrams (P&IDs).
2 Develop a 3D representation of the technical system on the basis of information collected from P&IDs, functional descriptions, 2D pictures of equipment and operating procedures.
3 Create a VR prototype using appropriate technology (e.g., the Reviatech software) in order to simulate plant equipment, user interactions and appropriate control panels for using the equipment. At this stage, a decision should be made regarding the fidelity of the VR environment (i.e., immersive versus desktop VR).
4 Use VR prototype to carry out safety previews for lengthy operations ahead of performance.
5 Create training scenarios and collect performance to evaluate training effectiveness.

The first stage of risk assessment requires a description of the sequence of tasks, the use of equipment, the operator procedures and the conditions of work (e.g., fatigue, time pressure and workload) which can be provided by a method of task analysis. Table 10.1 shows an extract of the car tanker

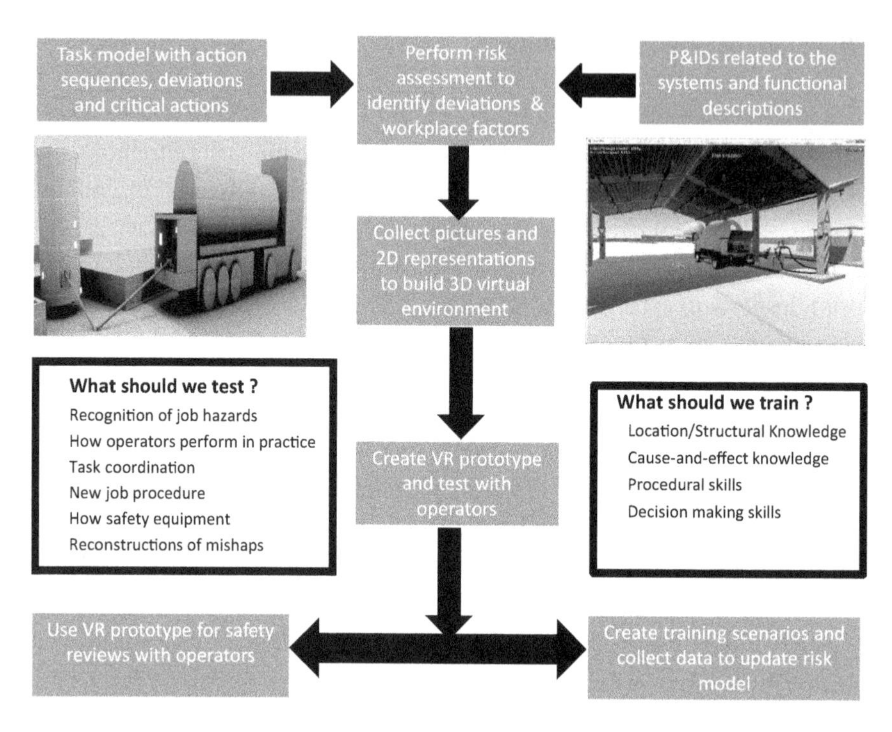

Figure 10.1 A risk-based methodology for VR training

loading procedure with some warnings and notes that could prevent the occurrence of human errors. The sequence of task steps and the warnings were derived from a collection of standard operating procedures, safety rules, and interviews with site operators and tanker drivers.

The second stage involves the modelling of the work environment. Several 3D models of the buildings and process equipment were created using photogrammetry – i.e., digital pictures of each building or discrete piece of equipment were taken from a variety of angles to create a map of the object. Software was then used to combine the pictures and extrapolate a 3D model. While not having the same accuracy as a CAD design model, the process was very fast and accurate enough for the simulation. Where required, many areas of the plant were pulled from Google Earth, which kept costs low while not compromising on the detail needed.

The third stage requires access to specialized VR software tools (e.g., rapid prototyping) in order to make a dynamic work environment that describes the behaviour of process parameters, the constraints on the use of equipment, the allowed operator actions and the feedback received by operators.

The virtual prototype begins with a basic 3D model of the workplace and the equipment, including the behaviour of the equipment and process. Feedback from real operators can be used to improve this model and the 3D model can be enriched with more details, such as material and textures. To minimize costs, desktop VR systems can be developed where the workplace can quite easily be 'drawn' from pictures, blueprints or photos to get a basic initial drawing. More details can be added to the specific points of interest, and these can be manipulated in the 3D environment, to get a feel for how the design will work. Changes can be made and assessed quickly, resulting in a collaborative, agile design environment. In the TOSCA project, rapid prototyping was based on a 3D editing software produced by Reviatech.

Figures 10.2–10.5 show pictures taken from the rapid prototyping tool that produced a desktop virtual environment for users to perform the car tanker loading scenario in a realistic way. Reviatech also developed an immersive VR

Table 10.1 Extract of task analysis with warnings in relation to human errors

1.	Connect car tank to the transfer station	Perform operations according to specified plan
2.	Park car tanker at the designated transfer location	Locations no. 1–4 available
3.	Connect the grounding cable to the tanker	*Do this before any other operation*
4.	Connect the flexible hose with break-away coupling	The connection must be made inside the car tanker's cabinet
5.	Open the manual transfer valve for media at the location	Valve selection depends on the location and gas choice. ***Warning: if valve closed and charging pump operates, the pressure could lead to hydraulic hammer and opening of PSV on the line – LPG release, fire hazard!***
6.	Open the manual input valve at the car tanker side	Operate at the transfer station – tanker ***Warning: if valve closed and charging pump operates, the pressure could lead to hydraulic hammer and opening of PSV on the line – LPG release, fire hazard!***
7.	Visually inspect the state of hose coupling for leaks	***Warning: if leak occurs anywhere on route, stop the transfer, inform manager!***

(*Continued*)

Table 10.1 (Continued)

8.	Go to the energy dispatch building and operate valves 15 and 14	Allow for some time to get to the building
9.	Go to compressor building and open valves 155 and 154	Allow for some time to get to the building
10.	LPG transfer into the car tanker	Perform operations according to specified plan
11.	Monitor the flow, pressure and amount of product transferred	Amount of product can be determined by measuring mass transfer *Warning: in a case of propane transfer at high temperatures, the total pressure in car tanker could reach the PSV opening set value (16.7 bar) leading to vapour phase release at the top – fire hazard! Monitor pressure in tanker!!! Stop transfer, remove ignition sources, inform manager, start cooling of the storage spheres.*
12.	Monitor tank level via built-in level indicator device	Occasionally during transfer; stop at or less than 80%!
13.	Stop transfer manually at planned amount or max. 80% (by local red STOP button)	Operation at the transfer station ***Warning: in case of overfill over 80% stop transfer, inform manager, unload tanker!***
14.	Stop manually if emergency occurs (by local red STOP button)	*In case of emergency stop transfer, inform manager, activate fire alarm!*

mode where operators could use special goggles and a mobile telephone to store the VR model; however, the immersive VR mode was not tested in this case study. Figure 10.2 shows a panoramic view of the car tanker and the surroundings where users can navigate with a joystick. The virtual panel through which users perceive process parameters and manipulate controls is presented in Figure 10.3. On the installation side, Figure 10.4 shows the arrangement of valves and pumps in the vicinity of the car tanker that are necessary for the transfer of product form the storage tanks. In addition, Figure 10.5 shows the firefighting equipment and the systems for emergency response.

The desktop VR system did not allow operators to handle directly process equipment but allowed them to use a joystick and perform all manipulations. This implies that users should also practice their tasks on real equipment to

Figure 10.2 Panoramic view of the car tanker and surroundings

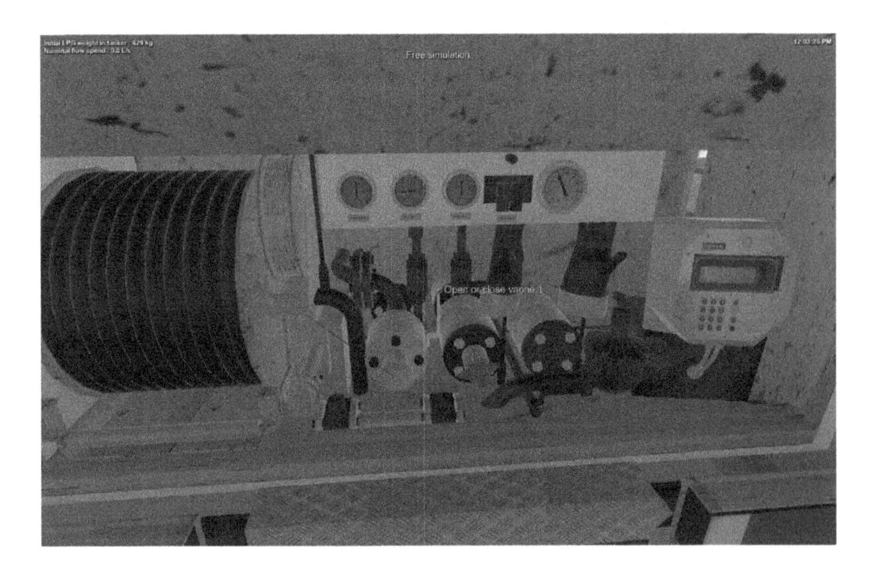

Figure 10.3 The main control panel of the car tanker

get a feel for the process. Nevertheless, the desktop VR system allows users to practice complex procedures in loading the car tanker. The behaviour of process parameters (e.g., flows, pressures and levels) was specified in a simple plant model that comprised a number of rules and static diagrams, describing the relationships between the parameters and their response to

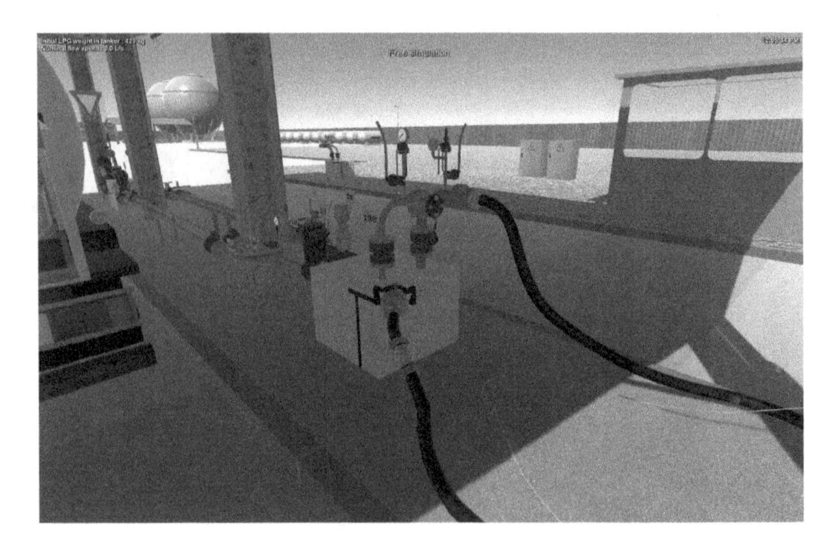

Figure 10.4 Layout of pipes and valves

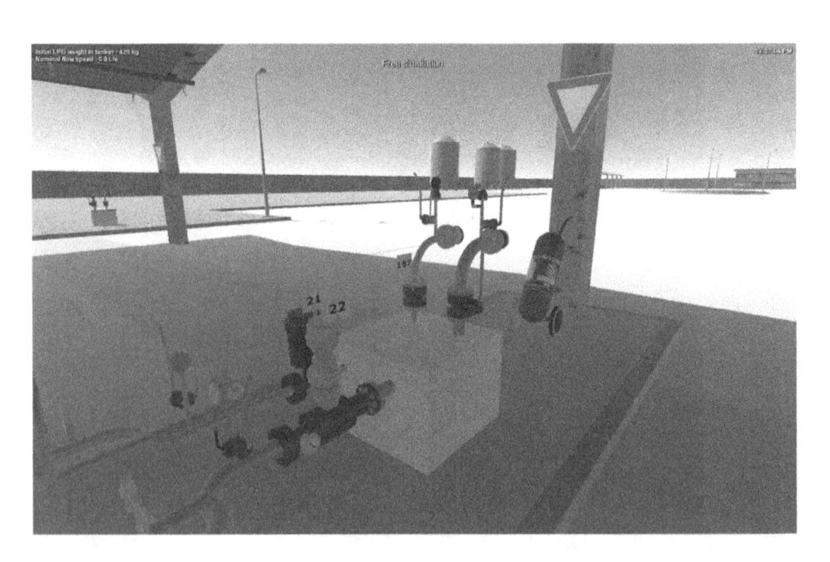

Figure 10.5 Firefighting equipment

their manipulation by the users. Although a dynamic simulator of the system would increase the fidelity of the technical system, a static plant model was perceived as acceptable for managing normal scenarios in the transport of the product. It is anticipated that more complex scenarios that deal with the management of process failures and instrumentation faults may require a more complex plant model.

However, it was possible to introduce some sort of variability into the operations that required users to transfer their procedural skills to a variety of situations. For example, it was possible to manipulate several system features such as, availability of people and equipment, weather conditions and failures of the tank level indicator. By performing many simulations, each with different settings, it was possible to record the responses of the users that were ineffective and make recommendations for changes in the design of the system or the training of users.

The VR environment can be used either for performing safety previews of critical tasks or for developing operator training programs. Safety previews are very useful when supervisors want to foresee alternative ways of doing a complex task, identify likely job hazards, warn operators of likely errors and try out several re-allocations of tasks to different operators. The VR system allows a large group of supervisors and production personnel to assemble in the same place and oversee how a complex procedure can play out in the future. Personnel who are unfamiliar with certain equipment and procedures can get an overview of the technical system fairly quickly and participate in the re-design of a critical procedure. The safety review can identify critical tasks that should be supported in terms of procedural guidance or team supervision. An alternative type of support would be operator training.

The design of training requires that some sort of Training Needs Analysis (TNA) has been performed in order to specify what skills should be learned, how tasks should be practiced, what virtual reality features and simulation systems are required and how training should be delivered. To master the process control skills to the required performance standards, adequate practice should be allowed in the virtual environment and preferably under a variety of situations. Finally, the performance of training should be assessed on several criteria, including: the speed of executing a task, the task accuracy and the number and types of errors made.

It seems that the Training Needs Analysis (TNA) stage is the most important stage of the training system and deserves further attention. First, the TNA should identify the tasks for training and the required types of skills and knowledge (e.g., procedural skills and cause-effect knowledge). Second, the features of virtual environment (e.g., multiple sensory cues, multiple perspectives, and immersive 3-D representations) are likely to influence both the *learning experience* (e.g., the information that learners attend and the strategies they use) and the *learning outcome* (e.g., the skills we learn, the types of errors made and our ability to remember tasks). An important issue here is the representation of the technical system in terms of dynamic features. For instance, a static system will not allow trainees to perform the tasks in ways that differ from standard operating procedures. In contrast, a task simulation would allow people to explore different ways of doing their tasks. As a result these factors will influence the modes of human interaction with the virtual reality (e.g., animations, procedural mode, free mode of interaction). Third, the training needs analysis should identify the training methods to be used for different types of skills, the variety of scenarios to be used (e.g., normal scenarios, emergency response scenarios etc.) and the number of training exercises.

Validation of VR training

The VR system allowed training of operators in two interaction modes:

- **Guided mode of interaction** where operators were not allowed to deviate from the normal operating procedure.
- **Free mode of interaction** where operators were free to manipulate the valves and pumps in their best choice.

It is worth noting that in cases where a critical error occurred, the system provided operators with critical warnings as well as offered the possibility to start again from the point where the error was made. In this way, the operators did not have to start all over again at the part of the procedure that was correctly executed.

The validation of the case study was done in terms of a training session with seven operators and a safety manager. A questionnaire was designed specifically to address the following aspects of VR training:

- Navigation in the VR environment and interaction with the plant equipment and the control panel
- Representation of hazards and fidelity of hazards in the VR world
- Control of the scenario by the trainees – e.g. backtracking, speeding up the process, hiding or uncovering instruments and complying to warnings
- Assessment of trainees, performance measures (e.g. speed, errors, checks), facilities for recording the session etc.

The questionnaire that was used is presented in Table 10.2 for the four areas of evaluation.

Table 10.2 Training assessment questionnaire

Navigation and interactions • Is there a realistic representation of the 3D environment of the car truck and its surroundings? • Is it easy to navigate in the 3D environment? (e.g. find where you are, move to other places etc.) • Is it important to increase the fidelity of the environment so that you are able to manipulate valves and hoses instead of just clicking on the icons? • Is it acceptable to use low fidelity representations for some tasks such as the manipulation of valves inside the energy building etc.? • Is it important to visualize critical parameters that are usually invisible? • Is the adjustment of zooming IN and OUT satisfactory?	**Control of the scenario** • Does the system allow you to backtrack to a previous step when you are not happy with what you did previously? • Is it important for you to see any warnings before a critical action in order to avoid certain errors? • Is it important to present new information that is not normally available on the job? • How important is it to be able to see the temperature and pressure inside the car tank? • Can you practice any 'WHAT-IF' exercises in the scenario and see the results? • What actions or tests were not possible to perform in the VR system? • Is it useful to be able to go faster in the scenario for routine phases of work?
Hazards • List the hazards and 3D visualizations that exist in the VR system (e.g., worn-out hoses, inoperative instruments, miscalibrated instruments, hot temperatures that affect tank pressure, ice on the ground, etc.) • Does the VR system allows you to spot any hazards related to sparks, static electricity, hot surfaces, welding etc.? • What other hazards would you like to have simulated as well? • Is it important to see visualizations of product leakages from the hoses, valves and car tank in terms of clouds of vapour or liquid pools?	**Trainee performance** • Does the system record the time that it takes to do the job? • Does the system record the errors that have been made? • Is it important to record the checks made by the operators? Is it important to be able to record each session and the errors made by others? • Is it a good training method to record errors of others and ask new trainees to detect what was done wrong? • Does the guided mode help you to get familiar with the system and perform later in the free mode? • Have you learned something that was not clear from the actual job? • What are the unique benefits of VR that are difficult to obtain in real life?

Overall, the operators were satisfied with the quality of training and thought that VR can be an efficient means of delivering training for process control skills. Some operators commented that they would like to practice with the full version of the plant model in order to get a more accurate reading of the temperature and pressure parameters of the truck vessel. Other comments related to the possibility to simulate a wide range of tasks regarding the handling of emergency scenarios such as the spilling and cleaning of product as well as the management of fires.

The Plinarna Maribor's management had identified two safety issues that could be used to test the capability of virtual reality and allowed us to examine the extent that they have been fulfilled. First, the interviews and responses to the questionnaire clearly confirmed the usefulness of safety previews of tasks within a team environment in order to visualize hazards before the job is done. The second issue related to the training of operators to recognize hazards and manage a procedural task. The experienced operators from the site commented that the VR environment was directly useful for the training of novice operators in a safe way. Furthermore, the logging of the achieved performance allowed an unbiased evaluation and recording of human tasks[1] under VR conditions before they were practised in the real operational environment.

Transfer of results

The VR training system can be transferred to similar contexts of operation regarding the transportation of dangerous materials, the transfer of products between storage tanks, the filling of vessels etc. The same system could be applied to the KNE case study (see Chapter 9) with minor modifications only.

The transferability of the VR system to other plant areas could be increased by developing a library of process equipment with their relevant job procedures. In this way, Small to Medium sized Enterprise (SMEs) would be able to build a range of process equipment for training operators in several process control tasks. The library of virtual items usually contains generic process equipment that needs to be tailored to the operating characteristics of the particular plant site. A VR system with a library of virtual plant items can become economically accessible to SMEs when several companies from the same industrial domain form a coalition that shares the initial investment.

With a further small cost, the VR library can be tailored to the needs of a particular SME.

The present VR prototype was a high fidelity simulation of process equipment although the plant simulator was not very sophisticated. It is anticipated that equipment fidelity could be lowered to achieve a more economical solution without any detrimental effects to the quality of training. In this case study, it was possible to run a scientific experiment to determine whether VR training provided a considerable advantage to other traditional training and test the performance of operators in the real world. Nevertheless, VR training should be supplemented with real world training so that operators receive adequate practice with the real equipment.

Conclusions

The VR training system is not an 'off-the-shelf' solution that remains independent from the other TSM methodologies. The VR training system uses a training methodology that is based on other risk assessment processes in the TSM framework. It makes use of task analysis and hazard identification methods to screen critical tasks to be considered as part of Training Needs Analysis (TNA). In other words, this is a risk-based training system which can be used to make decisions before training is delivered. For instance, if the risk assessment provides evidence that new safety equipment should be introduced or new control panels should be designed to minimize risks, then this should be respected. After the implementation of the new safety measures, the tasks should be re-appraised and the organization should consider the scope for improving the skills of the operators. Hence, SMEs that wish to apply new information technologies for training are better off than others when they use VR training systems. This is because part of the whole effort in safety training has already been done as part of other activities such as risk analysis.

VR technologies are becoming more cost-effective, particularly for organizations that already have a 3D model of their plant or equipment that can be easily transferred and used as a basis of a VR environment. This means that this vision of safety management will soon be achievable, even for small organizations. The technology gives us the ability to test out new ideas for small or large changes to processes and procedures; for example, creating a virtual model of a new production line and asking operators to adjust it to their work, suggesting new features or removing elements that get in their way. Also maintenance staff running through their planned maintenance work could identify any problems that might prevent their work from being completed on time. VR allows these kinds of tests to be run in a risk-free environment at low cost. The transfer of the results from the VR tests to the real environment will be invaluable to organizations seeking to minimize the impact of changes to production and safety.

Note

1 Specifically, operators asked if real time and action logging features could be enhanced by automatic video recording of the trainee's performance.

References

Aggarwal, R, Black, SA, Hance, JR, Darzi, A & Cheshire, NJ 2006, 'Virtual reality simulation training can improve inexperienced surgeons' endovascular skills. *European Journal of Vascular and Endovascular Surgery*, vol. 31, pp. 588–593.

Balfe, N, Leva, C & Sbaouni, M 2014, 'Picture this', *The Chemical Engineer*, no. 881, pp. 32–5.

Colombo, S & Golzio, L 2016, 'The plant simulator as viable means to prevent and manage risk through competencies management: Experiment results', *Safety Science*, vol. 84, pp. 46–56.

Cosman, PH, Cregan, PC, Martin, CJ & Cartmill, JA 2002, 'Virtual reality simulators: Current status in acquisition and assessment of surgical skills', *ANZ Journal of Surgery*, vol. 72, no. 1, pp. 30–4.

Gopher, D 2012, 'Skill training in virtual reality environments: The calls, challenges and prospects of a new domain', in M Bergemasco, B Bardy & D Gopher (eds.), *Skill training in multimodal virtual environments*, Taylor and Francis CRC press, Boca Raton.

Gutierrez, T, Gavish, N, Webel, S, Rodriguez, J & Tecchia, F 2012, 'Training platforms for industrial maintenance and assembly', in M Bergamasco, BG Bardy & D Gopher (eds.), *Skills training in multi-modal virtual environments* (pp. 227–39), CRC Press Taylor & Francis, Boca Raton.

Leva, MC, Naghdali, F, Balfe, N, Gerbec, M & Demichela, M 2015, 'Remote risk assessment: A case study using SCOPE software', *Chemical Engineering Transactions*, vol. 43.

Lin, F, Lan, Y, Duffy, V & Su, A 2002, 'Developing virtual environments for industrial training: Information sciences–informatics and computer science', *International Journal*, vol. 140, no. 1, pp. 153–70.

Nathanael, D, Mosialos, S, Vosniakos, G & Tsagkas, V 2016, 'Development and evaluation of a virtual reality training system based on cognitive task analysis: The case of CNC tool length offsetting', *Human Factors and Ergonomics in Manufacturing & Service Industries*, vol. 26, no. 1, pp. 52–67.

Sebok, A, Nystad, E & Droivoldsmo, A 2002, 'Improving safety and human performance in maintenance and outage planning through virtual reality-based training systems', in *Proceedings of the IEEE 7th Conference on Human Factors and Power Plants*, Scottsdale, AZ.

Seymour, NE, Gallagher, AG, Roman, SA, O'Brien, MK, Bansal, VK, Andersen, DK & Satava, RM 2002, 'Virtual reality training improves operating room performance: Results of a randomized, double-blinded study', *Annals of Surgery*, vol. 236, no. 4, p. 458.

Tate, DL, Sibert, L & King, T 1997, 'Using virtual environments to train firefighters', *IEEE Computer Graphics and Applications*, vol. 17, no. 6, pp. 23–29.

Vankipuram, M, Kahol, K, McLaren, A & Panchanathan, S 2010, 'A virtual reality simulator for orthopedic basic skills: A design and validation study', *Journal of Biomedical Informatics*, vol. 43, pp. 661–8.

Total project planning

Integration of task analysis, safety analysis and optimisation techniques

Maria Chiara Leva, Nora Balfe, Costanza Ciarapica-Alunni, Steve Prast and Stephen O'Mahony

Introduction

The safe and successful completion of maintenance and overhaul procedures is dependent upon the collaboration of different departments and individuals, the clear planning of the work and the availability of the required resources. For complex or rarely performed procedures, the competence and knowledge needed for planning and mitigating the risks associated with the project may be spread across different parts of the organisation. However, accessing and utilising this knowledge is critical for de-risking major projects and investments. Major projects, for example the delivery of the London Olympic Park, are increasingly placing value on the ability of suppliers and contractors to deliver projects with the highest levels of safety (Shiplee et al. 2011), and the criticality of safe performance of maintenance procedures is illustrated by major accidents such as the Piper Alpha, the Clapham Rail Disaster and the Texas Oil Refinery. In a study of 183 process industry major accidents, Okoh and Haugen (2014) found that 44% had a link to maintenance and of these, deficient planning/scheduling/fault diagnosis were a cause in; while maintenance is also considered a key factor in approx 20% of all occupational accidents, and 10–15% of all fatal accidents (OSHA 2011). The need for safer maintenance must also be balanced with the business requirements for time and cost effective completion of maintenance activities. This chapter presents a methodology for the elicitation of the information required to fully plan a maintenance activity, assessing both safety and efficiency goals. The methodology is based on a participatory approach that harnesses the existing knowledge in the organisation and engages key stakeholders in planning activities to help ensure safe and timely completion of activities. This chapter also shows how it was concretely applied to a case study in a power plant (Balfe et al. 2017).

Integrated methodology

An outline of the methodology proposed for total project planning is reported in Figure 1.1, it can be broken down in the following steps:

1 An initial participatory workshop uses a first version of the project plan (if available) to structure the workshop and elicit detail on the tasks to be performed from beginning to end from all participants.
2 The outputs of the workshop are used to construct a detailed task analysis, forming the basis of the updated plan. This is then used to structure a second participatory workshop, identifying the key risks for each phase of the work and generating mitigation actions in collaboration with the participants. A further update of the plan is made after this workshop to make any adjustments required as part of the mitigations.
3 The second version of the plan is then used to support the Monte Carlo analysis, which can be used to identify critical activities and examine the impact of varying resource levels.

The participatory approach is designed to engage stakeholders from across the organisation in the planning, and ultimately the successful completion, of the maintenance project. This approach draws strongly on the area of Participatory Ergonomics, an approach concerning "the involvement of people in planning and controlling a significant amount of their own work activities, with sufficient knowledge and power to influence both processes and outcomes in order to achieve desirable goals" (Wilson 1995, p. 1071).

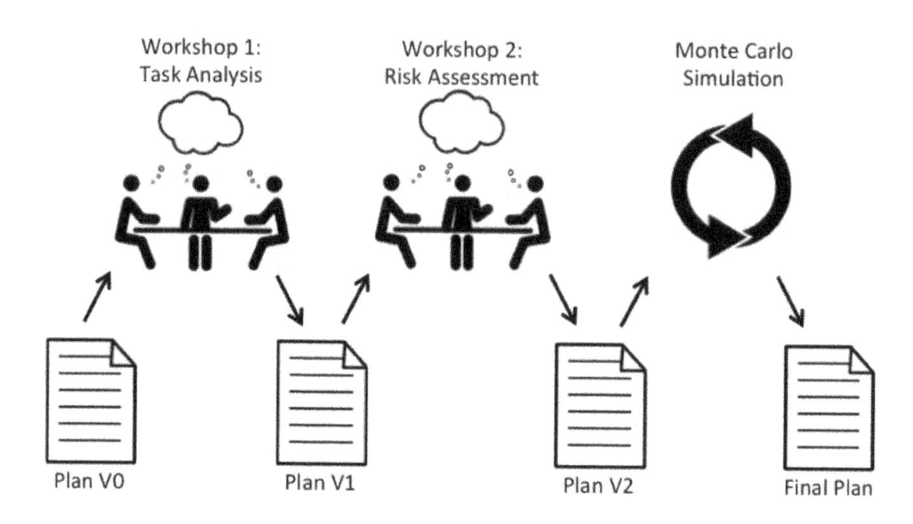

Figure 11.1 Participatory Planning Methodology

The data collection and analysis is based on bottom-up estimation techniques (PMI 2013) captured in a tabular task analysis that is subsequently used as the basis of the risk assessment. Task analysis is the human factor equivalent of a functional analysis in systems engineering. Functional analysis is a description of system functions achieved by arranging functions in logical sequences. Decomposition of higher-level functions adds more detail and allows system objectives to be allocated to lower-level functions. Similarly, task analysis identifies and examines the tasks performed by human operators when interacting with the system (Kirwan & Ainsworth 1992).

The task analysis is used as a baseline for a Hazard Identification Study (HAZID). The hazard identification is based on screening the activities considering five types of domain specific threats/prompts to identify possible issues connected with each step of the task analysis using a structured workshop format similar to a HAZOP study (Kletz 2006) where the nodes are individual (sub)tasks instead of parts of a plant. The approach can be referred to as TASK HAZID (Leva et al. 2012; Gerbec et al. 2017). The Hazid study involved a team workshop exercise with the team having the appropriate expertise, including knowledge of the plant and the planned operation, safety analysis, and human factors knowledge.

The five main domain specific threats/prompts used were:

- Logistical issues: lack of resources (manpower, cranes, equipment), unavailability of required parts, damage to required parts and/or

equipment, space constraints, improper provision for access ways and laydown areas, undetected faults
- Mechanical hazards: E.g. moving parts and materials, high pressure fluids, fragment and liquid metal projection, FODs left in after repairs, noise, vibrations, improper weight support, quality of welds
- Automation issues: miscalibration, undetected faults, etc.
- Process safety hazards: dangerous substances, hot surfaces, loss of containment, fire, *atmospheres* that are potentially *explosive* (ATEX), interlocks being disabled and not reset
- Electrical hazards: improper isolation, faulty connections, inadequate parts and fittings, arc flash
- Others: Human factors, ergonomics and mechanical lift supports, confined spaces, external weather conditions or microclimate considerations

The issues identified through the TASK HAZID were ranked using a risk matrix to discriminate between different safety and/or productivity impacts. That involved assigning the consequence classes on a 1 to 5 point scale, and similarly estimating their likelihood of occurrence, so as to determine their correspondent risk values.

The consequences and likelihood of each deviation were screened, and mitigations recommended where appropriate.

Finally, a Monte Carlo optimisation based on discrete-event simulation was used to examine the possible impact of different resource configurations. Discrete-event simulation models a process as a discrete sequence of well-defined events in time. Such events occur at a particular instance in time, marking a state change in the process (Robinson 2004). Discrete-event processes must include predetermined starting and ending points, and a list of discrete events that occur in between these points. Discrete-event simulation is commonly used to monitor or predict procedures and processes in various industries, such as manufacturing. The final aim of this simulation is to define a precise scheduling of the listed tasks, considering all the external and internal constraints characterising the whole activity to be organised. In order to provide a highly reliable plan of the tasks, the simulation should take into account a number of variables to consistently adhere to reality. Especially when the activity is characterised by a high level of internal and/or external constraints, it is of extreme importance to include the uncertainty affecting the hypothesis used to introduce those constraints in the simulation. In the present case study, the main constraints refer to the mutual conditioning of some subtasks (i.e. one task cannot be performed unless another one is already complete), availability of resources (e.g. people and tools) and the time of some external events necessary for accomplishing the procedure in safe conditions (i.e. time of tides). In this sense, the Monte Carlo (MC) method has been identified as a suitable tool to run a discrete-event simulation under uncertainty since it allows the computation of a mean value and a variance of the given quantity under investigation. This quantity is governed by a known phenomenon depending on a set of variables characterised by a level of uncertainty (i.e. the input variables are introduced as a set of mean values and related variances or in the form of Probability Density Function, instead of a set of definite values). MC computes a set of estimations of the final quantity based on different values of input variables generated according their Probability Density Function. The final output of the simulation will be an average of the quantity under investigation and a related variance depending on the variance of the input variables. In this specific case, the exploitation of MC is very advantageous for estimating the final planning of an activity taking into account the uncertainty declared by the stakeholders about the duration of each task. Indeed, the application of the Monte Carlo optimisation gives a more nuanced result from the analysis than is possible through the task and safety analysis. It allows different options to be tested and the expected impact on the project plan to be evaluated.

Case study: cold water system overhaul

The developed methodology was trialled within an electricity-generating organisation in Ireland. The organisation regularly undertakes maintenance and overhaul procedures during unit outages at their generating stations. These complex works have significant safety, productivity and ultimately financial consequences if not successfully completed; however, detailed project management is difficult due to lack of firm planning data, variation in the work undertaken in each overhaul, the different departments involved in the works, time pressures associated with normal running of the plant and the focus of engineering teams on technical planning.

For this case study, the overhaul of a cold water system during two overlapping unit outages was considered. The planned works involved the replacement of two valves and the inspection and repair of the busmain. The work must be undertaken during the outage of two of the three generating units, so that reduced cooling water flow is required for the overall station, enabling the isolation of the necessary parts of the cold water cooling system. The outage of both units was initially planned for five days, leaving a short time window in which to complete the work, and significant losses to the plant would be incurred through lost revenues if start-up were delayed. The time pressure associated with the works also increases the likelihood of errors that may impact both safe and timely completion of the work. The aim of this case study was to provide a project plan for the overhaul procedure, based on the participatory approach to engage all stakeholders and integrate that plan with a risk assessment of the works covering safety, cost and time. The resulting plan should account for individual tasks, consider access to and use of resources and identify the main risks, in terms of operational delays and process safety, capable of undermining the successful outcome of the project.

The first participatory workshop was held approximately 2.5 months prior to the planned start date of the works. This workshop detailed the task analysis and also documented any risks and mitigations that emerged in the course of the discussion, although these were not the focus of this workshop. Additional one-to-one meetings were held with key stakeholders over the following months to clarify some remaining points and the plan was documented in the form of the task analysis and Gantt and PERT charts. The second workshop was held one month prior to the works to finalise the risk assessment. The Monte Carlo simulation was finalised approximately two weeks ahead of the works start date.

The initial workshop was designed to engage key stakeholders and elicit the information needed for detailed planning. A basic scale model of the CW system was used to structure the discussion (Figure 11.2). Participants

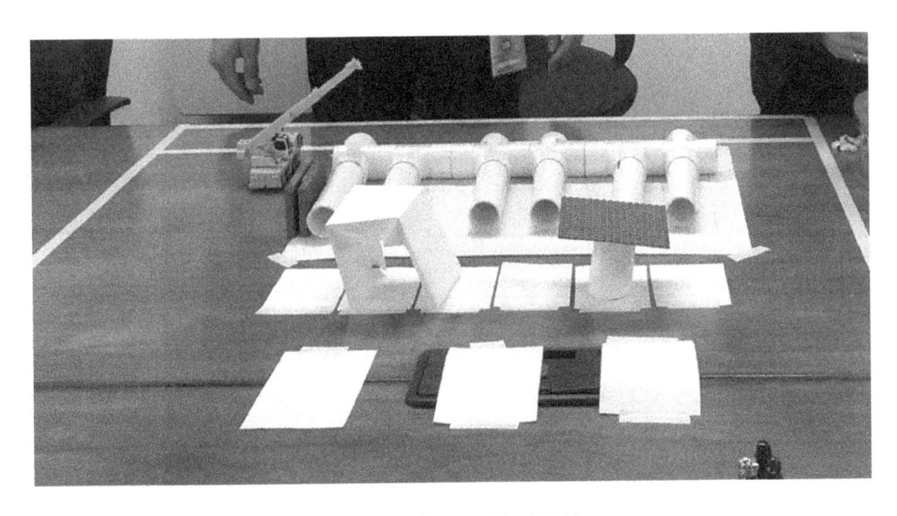

Figure 11.2 Scale model of CW system

were asked to use small figurines and toy cranes to represent the progression of the tasks on the model. Each participant was assigned one or more roles. This structured and immersive approach helped to elicit and document each task needing to be completed during the works.

A full list of tasks, from when the second unit shut down through to handing back of the CW system to operations, was elicited in the workshop and annotated with actors, tools and equipment and expected durations against each. Upper and lower limits of the expected durations were also noted for the future optimisation. The data collected from the workshop was structured in an Excel-based tabular task analysis, highlighting questions for further analysis and resolution. These were addressed offline in discussion with key stakeholders.

This process resulted in a detailed plan for the works, which was represented in both PERT and Gantt Charts. This plan was used as the basis for a hazard identification workshop with the same group of stakeholders considering likely hazards as a prompt to identify all the possible risks associated with each task so as to verify if any additional countermeasures may be necessary. The impact of each identified risk was considered in terms of Cost, Time and Safety and rated according to its perceived likelihood and severity, using the following scales (Table 11.1 and Table 11.2).

The risk matrix (Table 11.3) was used to calculate the overall risk score, by cross-referencing the assessed likelihood and severity. The red indicates high risks that must be addressed with actions to reduce the risk. Amber risks should be reduced if practical, while green risks can be accepted, although any further risk reduction measures that can be implemented easily and

Table 11.1 Likelihood table

Likelihood			
Rating	Name	Description	Quantification
1	Unlikely	Could happen but never heard of in the industry	$< 10^{-6}$ ev/yr
2	Remote	Has occurred in the industry	10^{-6}–10^{-4} ev/yr
3	Possible	Has occurred in the company	10^{-4}–10^{-1} ev/yr
4	Probable	Has occurred several times a year in the company	10^{-1}–1 ev/yr
5	Frequent	Has occurred several times a year in the location	>1ev/yr

Table 11.2 Severity table

Consequences			
Safety		Cost	Time
1	Minor injury	<10k €	< 3 hr delay
2	1–2 day LTA	>10k €	< 8 hr delay
3	Serious injury	>50k €	< 24 hr delay
4	1 fatality, permanent incapacity	>200k €	< 1 week delay
5	Multiple fatalities	>1M €	> 1 week delay

Table 11.3 Risk matrix

		Risk				
Likelihood	5	5	10	15	20	25
	4	4	8	12	16	20
	3	3	6	9	12	15
	2	2	4	6	9	10
	1	1	2	3	4	5
		1	2	3	4	5
		Severity				

cost-effectively should still be considered. This risk matrix, as well as the severity and frequency tables used, was aligned with those used more widely for safety management within the organisation.

Finally, an Excel-based Monte Carlo simulation was run on the basis of the task outline, the estimated durations and the required availability of resources, using a triangular distribution to describe the expected, upper and lower durations. Where upper and lower limits had been captured during the workshop, these were used; otherwise upper and lower limits were defined as 1.1 and 0.9 of the expected duration respectively.

Results

The workshop broke the works down into eight distinct phases of the revamping procedure:

Z. Preparation
A. River isolation

B. Draining and preparation of the cold water pipework
C. Removal of the valves (1/2 NRV and CW1/2 isolation valve)
D. Inspection and repairs
E. Installation of the new valves
F. Commissioning
G. Removal of river isolation

Figure 11.3 shows the PERT diagram for the project.

Under this current plan, the project was expected to be completed at 2330 on Tuesday, 11 August, or 87.5hrs (3 days, 15.5 hours) after the estimated start time of 0800 on Saturday, 8 August following the planned outage of two of the three power generation units in the station. The critical path is shown in light gray.

The PERT includes scheduled break times for those activities scheduled to occur during rest periods. These are reported in the duration boxes of the related activities underneath the main duration time. The plan is slowed by the need to wait for resources on three occasions:

1 C3: Only one crane is available and this is needed simultaneously to lift and hold both valves during their removal (C2, C3). Therefore, the removal of one valve will have to wait for the crane to become available. In the PERT, this valve is the NRV, putting the remaining activities involving the NRV on the critical path. If the NRV is instead removed first, the remaining isolation valve tasks will instead be on the critical path.
2 D4: The repair teams work only dayshifts, but to complete the repair of the isolation valve flange face would require working through the night. A wait time of 12 hours has been added to account for this time.
3 D5: Again, the repair team is not available to start work at 2330 (the scheduled removal time of the NRV), and this task must wait to begin at 0800.

The risk assessment based on the detailed plan identified 10 green risks and 18 amber risks. The green risks were deemed to be sufficiently well mitigated and were not further analysed. Examples of amber risks included:

- A delay in the operational isolation (Likelihood: 3, Severity: 3; Risk = 9). The operations team are likely to have a high workload after the unit comes off load and may not be able to provide the isolation in the required timeframe, causing a delay to the work. A dedicated operator,

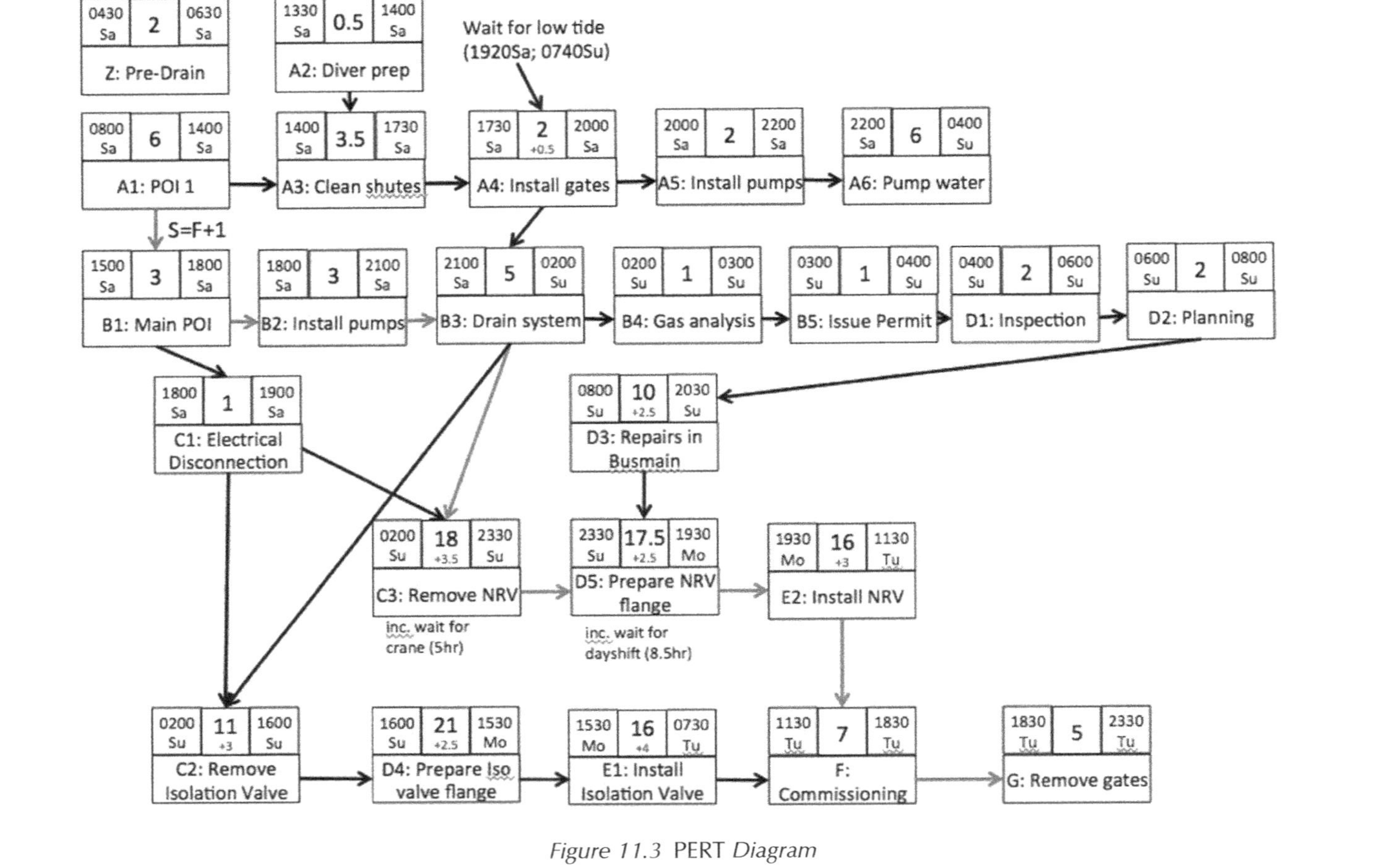

Figure 11.3 PERT *Diagram*

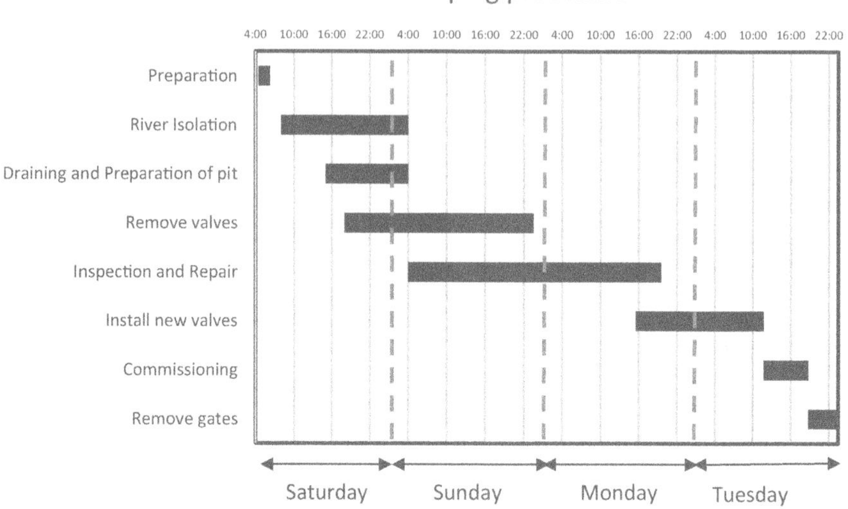

Figure 11.4 Gantt for overall operations

and clear planning of the works were the recommended control measures for this risk.

- Diver injury (L: 2, S: 4; R = 8). The cleaning of the chutes is a high-risk activity for the divers. The recruitment of an experienced diving team and careful monitoring of their planning and risk assessment should mitigate this risk.
- High winds prevent crane use (L: 3, S:2; R = 6). High winds may prevent the crane from dropping the river isolation gates into place, causing a short delay to the works until the winds die down. The only possible mitigation is to check weather conditions in advance of the works to determine the possible impact and whether to continue.
- Damage to existing pipework around the valve being removed (L: 3, S: 3; Risk = 9). The removal of the large valves using the crane holds the potential to damage the remaining pipework. Procedures for the careful removal of the valves were developed to manage this risk.
- Lifting hooks unable to hold valve weight (L: 2, S: 3; R = 6). The lifting hooks may have become degraded over the course of time and may not be able to take the full weight of the valves. Inspection of the lifting hooks prior to the works is recommended as is the use of a sling around the body of the valve.

- Inhalation of toxic gas formed during curing (L: 2, S: 3; Risk = 6). A toxic atmosphere may form in the pipes during the curing phase of the repair works. Force ventilation should be installed to mitigate this risk.
- Injury of operator during blasting operations (L: 2, S: 3; R = 6). Operator PPE should be available and in use to prevent injury.

A Monte Carlo simulation of the planned works was run, using the expected, maximum and minimum values for each task as collected in the initial workshop and follow-up visits. The simulation used had two main constraints on its realism. First, break times cannot be accounted for in the current simulation. This was dealt with by adding 3 hours to every 24 hours of the final expected time. Second, the simulation assumes all resources are available 24 hours, so cannot account for the repair teams only being available during the dayshift. Nevertheless, the simulation provided some useful results.

Averaging over 20 runs, the simulation expects the work to be completed at 2240 on Tuesday, 11 August. This is very close to the expected time as calculated in the PERT, and suggests that the simulation is reasonably accurate despite the constraints.

If a second crane were to be used, the averaged result of the simulation suggests that the time required for completion could decrease by 7 hours, anticipating the return to operation to 1530 on Tuesday, 11 August. The use of the third repair team makes little difference on its own, but the use of three repair teams and a second crane makes the completion time a full 12 hours earlier, at 1000 on Tuesday, 11 August.

Validation

The approach was validated against the actual works as completed. The primary objective of the study was to de-risk the project by identifying the tasks, resources and risks in detail and ensuring the necessary measures were in place to maximise the probability of safe and successful project completion.

The main criterion was satisfied, in that the project was successfully completed during the outage. However, this is not sufficient to validate the method and additional analysis of the planned versus actual project completion was performed to this end. Feedback was also collected from key stakeholders.

Duration comparison

The work was calculated to take approximately 4 days (3 days, 15.5 hours), but in actuality took 7 days. The change in overall timescale was ascribed to additional time becoming available for the outage, due to overrunning of other planned works, allowing the schedule to be relaxed. Key differences were a delay to the start of Phase B, not working 24 hour shifts, and not running major tasks in parallel. Table 11.4 describes the expected duration for each phase of works, compared with the actual duration (not including breaks) and Figure 11.5 and Figure 11.6 describe the overall timeline of the

Table 11.4 Expected vs. actual task durations

Task	Expected Duration	Actual Duration
A. River isolation	19.5 hrs (12.5–26hrs)	13.5hrs
B. Prep of pit	13 hrs (11–16hrs)	8 hrs
C. Remove valves	11 hrs (6–15.5hrs)	12 hrs
D. Inspection and repair	14 hrs (12–31hrs)	19 hrs
E. Installing new valves	13 hrs (6.5–16.5hrs)	13 hrs
F. Commissioning	7 hrs (5.5–11hrs)	9 hrs
G. Remove gates	2.5 hrs (2–5hrs)	5.5 hrs
Totals	80 hrs (56–121hrs)	80 hrs

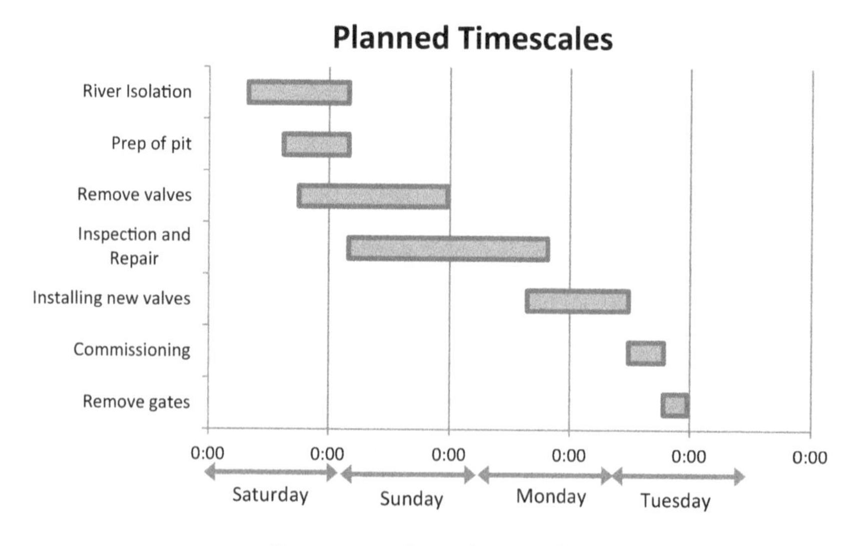

Figure 11.5 Planned timescales

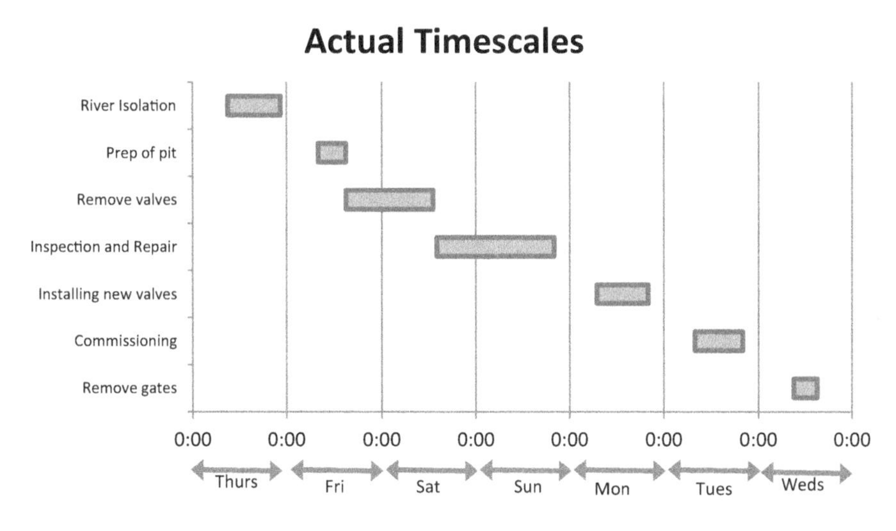

Figure 11.6 Actual timescales

planned and actual works respectively. The actual duration of the works exactly matched the expected duration, although this was likely a coincidence given the variation in individual tasks.

The majority of tasks were within the expected timeframe, with most tasks generally at the lower end of the expected duration. Phase B (prep of pit) was the only phase completed more quickly than the lower expected duration. This was due to a considerably faster drainage time of water from the pit than anticipated. However, the durations in Table 11.4 do not include an 18-hour delay to the start of the tasks, due to the need to wait for an isolation from operations. Phase C (remove valves) was completed in a time very close to the expected duration, however several of the tasks in Phase C had been planned in parallel, but were completed in series. The actual tasks of removing the individual valves were completed in a time closer to the lower expected limit (i.e. 5–6 hours actual duration per valve; expected durations were 8–10 hours, with a lower limit of 5 hours and upper limit of 13.5 hours). Phase D (inspection and repair) took longer than the expected duration, although still within the set limits. This was due to the completion of tasks in series that had been planned in parallel. Similarly with Phase E (installation of new valves) although the planned and actual durations are similar, in fact the tasks were completed close to the lower time estimate, but were completed in series rather than parallel, as had been planned. Phase F (commissioning) took slightly longer than the expected duration, due to the removal of the isolation. Finally, Phase G (remove gates) exceeded the upper expected duration. This was due to longer than expected times to remove the pumps from the system and to equalise water pressure with the river.

A final note on the accuracy of the estimation method is to compare the bottom-up estimate used to an initial top-down estimate prepared by the project leader. This estimated the works to take 11 days, a vast overestimation in comparison with the eventual bottom-up estimation, and an overestimation even compared to the relaxed pace of the actual project completion. While this 11-day estimate was an initial estimate and may reflect lack of accuracy at that point, it is also likely that the detailed decomposition facilitated a more accurate analysis of the final duration. This is interesting in the context of Besner and Hobbs (2004) finding that top-down estimating is more widely used the bottom-up estimating.

Overall, the durations were within tolerance, and the planned procedure provided a good estimate of the actual performance.

Risk analysis results

The risk analysis carried out in the TASK HAZID workshop identified 18 amber risks to the works. Additional risks were identified and mitigated during the initial task analysis, and in fact this approach is more integrated and appeared to offer a more natural way for stakeholders to consider the possible deviations and controls as they developed the plan. However, a drawback of the risk assessment based on the detailed task analysis was the fixation of stakeholders on individual deviations, rather than global risks to the project. These were under-represented, and key risks which did later manifest, such as the delay to the outage start time and the overrunning of another project during the outage (which in fact had positive consequences on the project), were not identified. The risk analysis aspect of the approach therefore needs to be re-considered, with perhaps a brainstorming session before or after the completion of the task analysis to identify wider risks to the project as well as risks to individual tasks within the project.

Cost benefit analysis

The planning and optimisation work conducted for the case study is estimated at one project manager (engineer) month of effort (i.e. 20 days), with a gross salary of €5,000. This includes the time of the various stakeholders (project manager, four engineers, operations supervisor, technician) for two workshops and an additional 6 days of effort by the project manager to analyse the data and generate the plan.

In contrast, an overrun of one day is estimated at €250,000 in lost earnings for the station, and the analysis helped ensure that the work could be completed within the available time window. The risks identified to the project calculated the potential risk exposure of up to €102,035, based on a multiplication of the frequency and severity scores determined in the risk workshop. The cost benefit analysis therefore clearly demonstrates the value of the pre-planning and analysis work in de-risking the project.

Feedback

Key stakeholders gave very positive feedback on the approach, with comments including:

> *I firmly believe that the attention to detail highlighted in your submissions prior to the project helped us to complete the work safely and without delay.*
>
> *Input made a huge difference and the CW was one of the jobs during our recent outage that had a very successful outcome.*

Specific benefits of the approach were:

- Early engagement and motivation of stakeholders across the plant, ensuring tasks and activities were clear and achievable during the works
- Clear planning that identified pre-work and resources required to de-risk the works
- Confidence on what was possible in the time window available, enabling re-planning decisions to be made smoothly during the outage when required.

Discussion and conclusions

This chapter has presented a methodology for the integration of safety analysis with project planning for maintenance procedures. The need for such a methodology is clearly demonstrated by the high number of process and occupational safety accidents occurring during maintenance activities, and the contribution of poor planning to those accidents (Okoh & Haugen 2014). The approach presented here aims to engage key stakeholders through participatory workshops, to capture their knowledge and expertise in developing a plan and to promote coordination between different stakeholder groups. Capturing the knowledge elicited in the workshops in a project plan allows the key issues to emerge and different options to be tested ahead of the works.

The case study presented in this chapter represents a complex maintenance procedure involving several different stakeholder groups and operating under tight constraints. The workshop approach was very successful in engaging the plant personnel, and allowed key efficiency and safety risks to be identified, discussed and mitigated ahead of the works. Although the eventual works did not follow the plan exactly, the clear documentation of the tasks and risks allowed the maintenance team to adapt to changing circumstances with confidence.

It was not possible to include all stakeholders in the workshops; in particular it was difficult to access contractor staff in this context and this is likely to be the case in many organisations. This limitation was overcome by documenting assumptions or questions for the contractor staff, and directing these queries to the relevant contractors after the workshops. However, it would have been preferable to better engage contractors directly in the whole planning procedure. A second limitation was the linear nature of the case study followed, which limited the potential benefits of the optimisation methods. Future research should consider less linear projects in order to demonstrate the additional benefits of optimisation. The MC analysis should also be improved to account for break patterns and to incorporate the risks identified.

Overall, the application of the integrated methodology in the case studies described in this paper has demonstrated the potential benefits in terms of increased safety, cost savings and time savings. The individual methods described are all available to organisations at little or no additional cost, and the use of workshops with key stakeholders serves not only to elicit the necessary information for better project planning, but also to engage the stakeholders in the project, highlight the key risks and motivate stakeholders to work together to ensure project success. The approach was positively received and showed its ability to provide an accurate and detailed plan of proposed works. The methods used are a combination of tried and trusted project management, human factors and risk assessment methods, but their combination and use in a participatory manner provides a stronger result with better engagement from key stakeholders, resulting in increased motivation for a successful project conclusion.

References

Besner, C & Hobbs, B 2004, 'The perceived value and potential contribution of project management practices to project success', *Project Management Institute*, vol. 37, no. 3, pp. 37–48.

Balfe, N., Leva, M. C., Ciarapica-Alunni, C., & O'Mahoney, S. (2017). Total project planning: integration of task analysis, safety analysis and optimisation techniques. *Safety science*, 100, pp. 216–224.

Gerbec, M., Balfe, N., Leva, M. C., Prast, S., & Demichela, M. (2017). Design of Procedures for Rare, New or Complex Processes: Part 1–An iterative risk-based approach and case study. *Safety science,* 100, pp. 195–202.

Kirwan, B & Ainsworth, LK 1992, *A guide to task analysis*, Taylor & Francis, London.

Kletz, T 2006, 'HAZOP and HAZAN, identifying and assessing process industry Hazards', *The Institution of Chemical Engineers*.

Leva, MC, Pirani, R, De Michela, M & Clancy, P 2012, 'Human factors issues and the risk of high voltage equipment: Are standards sufficient to ensure safety by design?', *Chemical Engineering Transactions*, vol. 26.

Okoh, P & Haugen, S 2014, 'A study of maintenance-related major accident cases in the 21st century', *Process Safety and Environmental Protection*, vol. 92, no. 4, pp. 346–56.

OSHA 2011, *Healthy workplaces: A European campaign on safe maintenance*, European Agency for Safety and Health at Work, Brussels.

PMI 2013, *A guide to the Project Management Body of Knowledge (PMBOK)*, 5th edn., PMI, Pennsylvania.

Robinson, S 2004, *Simulation: The practice of model development and use*, Wiley, New York.

Shiplee, H, Waterman, L, Furniss, K, Seal, R & Jones, J 2011, 'Delivering London 2012: Health and safety', *Proceedings of the Institution of Civil Engineers: Civil Engineering*, vol. 164, no. 5, pp. 46–54.

Wilson, JR 1995, 'Ergonomics and participation', in JR Wilson & EN Corlett (eds.), *Evaluation of human work: A practical ergonomics methodology*, 2nd and revised edn, Taylor & Francis, London, UK.

Management of technical and organizational changes in major hazard industries

Marko Gerbec and David Levovnik

Introduction

Business organizations are the subject of numerous incentives to adapt/change, usually due to market pressures, statutory requirements, etc. While changes occur in any organization, and reorganizations are one of the main management's tools to manage/improve the organizational performance, an additional care shall apply in establishments on the subject of major accident hazards. In major hazard establishments, the changes might apply to its organization (e.g., policies, roles, duties, hierarchy, staffing, communication routes, etc.), or its production related operations (changes in technology/production procedures, equipment, process conditions, materials [raws, intermediates, products], etc.). Historically, there have been numerous major accidents that can be at least partly attributed to deficiencies introduced by the changes. For that purpose, the management of changes is a recognized part of the process of safety management systems in major hazard industries. For example, in the European Union, that was introduced as a top legislation directive in 1982 and is now part of the required formal safety management system within the current "Seveso III" directive (EC 2012), while in the United States, similar federal legislation was introduced in 1992 (US 1992). The main principle to be followed in management of any changes in organizations is that any changes shall be prior to their implementation first evaluated for potential safety implications and be implemented and documented only after they are approved. It should be also added that it is not only about development projects of any size, but those principles shall apply to all changes that shall alter the original or established organization, equipment or procedures/technology.

While there are available guidelines on the principles and details to be considered in the management of changes (CCPS 2007, 2008, 2013; CSChE 2004; HSE 2016; Sanders 2005), there are not many scientific papers on the topic. Of the existing ones (briefly listed in Gerbec 2016), it worth mentioning that Zwetsloot et al. (2007) found that in the studied cases the managers of the organizations reported difficulties in addressing the complex dimensions in decision-making with the changes – more precisely, decision fragmentation, undesirable side effects and ambiguity. Thus, the changes should be evaluated for the impacts in an integrated way, addressing all applicable management levels and managed aspects, considering internally both technical and organizational dimensions (Gerbec 2016). Such an approach shall ensure that the risk information gap does not occur among the stakeholders in a change and shall build on the concept of situational awareness/common operational picture (COP; Endsley, Bolte & Jones 2003; Seppanen et al. 2013; Balfe et al. 2016), as being the essential features of total safety management.

In that respect, this chapter will present the topic of management of change/ safety change management as understood under management science (field of economics), management system and safety management systems, conventional management of changes (safety implications of technical changes), recently proposed integrated safety change management, and finally, we will present the method and results for evaluation of organizational readiness (maturity) of safety change management in industrial organizations.

Understanding the changes

In the first subsection we will briefly explain understanding of the management view on the change management, which (as it will be presented) does not explicitly consider implications on the aspect of safety. In further sections, we will briefly explain the other side: the safety part of management systems and management of change.

Management view

When changes occur in a business organization, they can have an important effect not only on the organization but also on different stakeholders. To understand the influence that changes can have, a special field of study called change management has been formed. By its nature, change management is an integrating field. As Hayes (2010) points out, it builds on various perspectives on organizational functioning that arise from different fields of study, such as finance, marketing, strategic management, organizational behaviour, operations management and so on. Factors like globalization, technological advances, development of new organizational forms and competitive pressures can create both new opportunities and new threats at the same time. If they want to survive in the ever-changing business environment, organizations will have to address those factors adequately. Managers at all levels should not only be competent at identifying the need for change, but should also be able to implement it successfully.

When it comes to management of the organization, change is often regarded as one of the few constants. Even more, for Paton and McCalman (2008) change and management are synonymous. They compare it with a journey, which is impossible to undertake if you do not first address the purpose of the journey, the path you wish to take and with whom you wish to travel.

Cameron and Green (2012) suggest that the understanding of an organizational change should be based on the basics assumptions about how the organizations really work (we can see them as machines, political systems, organisms, etc.). Therefore, we can address organizational change with different models and approaches. Because there are many different models that build on different assumptions about organizations, there cannot be a single approach to understand and successfully manage the change. In order to effectively manage the change or consult on this matter, we have to be able to use appropriate models or approaches for every specific situation. We have to understand our assumptions about change management so that we can challenge them and study the possibilities that arise from different assumptions. We must also have in mind that change does not have only an organizational dimension, but also includes individual and team change. However, there are some basic characteristics when it comes to managing the change.

Usually, we see changes as responses to the outside business world. Those responses are regarded as external triggers. For example, environmental issues can drive the government to initiate an implementation of stricter legislation, which consequently affects the organization. In order to meet the legal requirements, the organization has to change its technological aspect of operations. At the same time, there are a number of factors that present internal triggers of change. The organization can start a redesign phase to introduce a new product line or launch a new marketing strategy. Another typical example of this type is change in work responsibilities in order to adapt to new organizational structure (Paton & McCalman 2008).

Managing the change process

Change is usually managed far less effectively than it should be, because managers that are responsible for its management do not address some critical aspects of the change process (Hayes 2010). A conceptual framework can be provided with eight basic steps in the change process. By their nature, the steps are generic and can be applied to various changes at different levels (see Figure 12.1):

- First, we have to recognize the need for change and start with the change process. We have to recognize internal or external circumstances that require a change. This is a process, which involves perception, interpretation and realization that changes can, when not managed adequately, result in undesired outcomes. Next, we have to decide who will be responsible for managing the change and where we should introduce the external change agent that can assist with the above-mentioned process.
- The second step involves revision of the present state and identification of the preferred future state. In practice, both activities are often intertwined. The decision about with which of them we should start the process depends on the nature of the change at hand. Focusing on the present state will result in goals that are more cautious. However, when the change will have to be more radical it would be better to start with the identification of desired, future state.
- Plan and prepare for the implementation is the crucial step in the process. The implementation should not be viewed as only a technical activity. It also needs to address people and their perception of the change. It is necessary to consider whether are they ready for change, to which extent they are accepting it and if the process in any way threatens them. Therefore, the planning on which method to use for implementation is important.
- In the implementation step, we have to refocus from planning to action and implement all things that were planned. Monitoring and control of the implementation process are also important to ensure that change

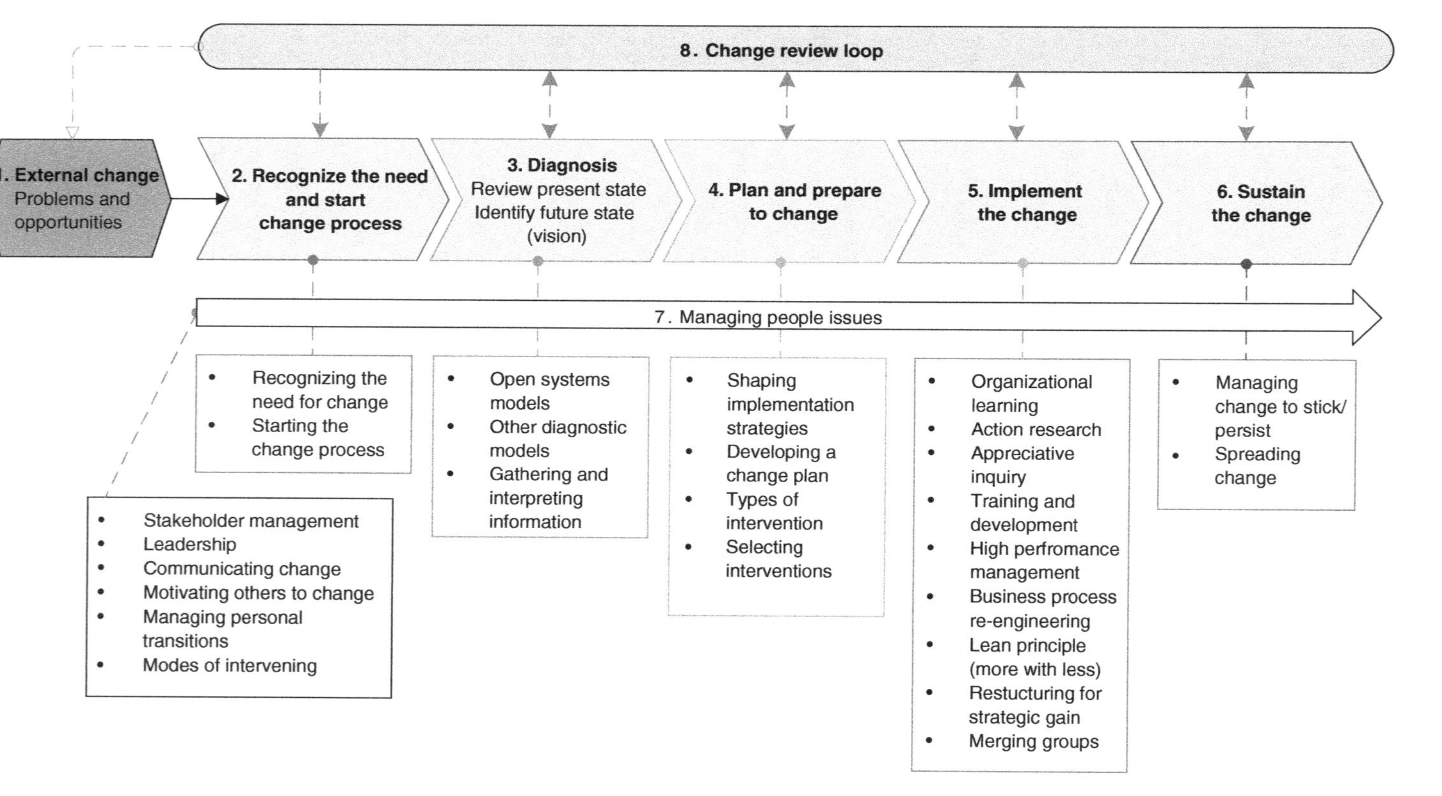

Figure 12.1 Key steps in change process (adopted from Hayes 2010)

is managed as intended. There are two basic methods of implementing change. If the nature of desired change is known (e.g. computerization or relocation of business process, etc.) it is relatively easy to manage the change according to predetermined process. In some other circumstances, it is not possible to determine the desired end stage of a change (e.g. losing market share, falling behind competitors, etc.). The goal may be broadly defined but the detailed specification of end state cannot be provided. In this case, we have to treat change as an iterative process.

- The next step is a review process that has to be more than just post-implementation activity. In practice, monitoring and review have to be ongoing activities. It often occurs that feedback is unavailable because the process of change management is too disconnected. Some managers may manage the change and fail to recognize the need to inform others that are involved in the process, or may address the problem without considering the impact of their actions. Even if managers are aware that others should be informed, there is always the question of who should be informed and if they will recognize the importance of the information.
- Sustaining change involves securing the gains of the change or in other words "making the change stick". To establish this requires feedback and reward systems that are able to monitor and reinforce the desired behaviour. However, there is also the ever-present need for updating the change.
- We also should not forget to manage the people. In addition to above-mentioned steps, change management must also address a broad variety of people issues. That may include stakeholder management, leadership, motivating people to change, communication, helping others to manage personal transitions and so on. Managers need to manage these issues at all steps of the change process and not just at the beginning when the strategy for implementation is shaped. The steps in the change process should not be treated as purely technical. There is often too little attention directed towards motivational issues related to the change.

One should also be aware that there is no universal approach to how to manage change which could be used in all situations. Therefore, the approach will have to be contextualized according to the individual case. Once again, we would like to point out that even though that change process suggests that change is a linear and rational process, this is rarely the case in practice (Hayes 2010).

Case of major change – mergers and acquisitions

Mergers and acquisitions are definitely one of the major changes in the life of the organization. There are different types of mergers that reflect different reasons why the organization chose this step. The main reasons are usually connected with growth, synergy, diversification, integration to achieve economic gains, defensive measures, or they can be a case of deal doing. Mergers have also one other important feature; they represent a change that in most cases goes wrong. According to literature, 70% to 80% of them fail. In such cases, acquisitions did not create wealth for the shareowners of the acquiring organization, as they should. The sustained losses can be devastating and in some cases even fatal for the organization (Cameron & Green 2012).

In the case of major change, like mergers and acquisitions, Cameron and Green (2012) suggest that change should be managed on all three dimensions (individual, team, organizational) in accordance with their specific needs. For the individuals, this type of change brings uncertainty and consequently also anxiety. They should know that things will be different; they have to understand what will change (what will be their new role, new responsibilities, etc.) and when the change will happen. A supportive environment for employees is also important, so they can try out different ways of doing things. We must also manage change at the team level. There, endings (end of a specific way of working, new team members, working environment, etc.), transitions from old to new (process of integration into new work environment) and new beginnings (set a clear new vision of the future work – putting up key performance indicators is not enough for most employee) can be most effectively addressed. At the organizational level, change process has to be selected in a way that matches the challenges of a specific situation. Here issues can be complex and diverse; therefore many different assumptions about change have to be combined.

Management systems

Organizations usually have to manage multiple aspects being important for their long term success and existence. While usually even the smallest organizations have some kind of formal management system documented in management rules and a set of standard (operating) procedures in order to preserve the knowledge, the complexity of the management system quickly increases with the size (e.g., number of employees) of the organization. As the basic functions of the organizations are more or less the same in all organizations, and for the purpose of facilitating the comparability among the various connected organizations (e.g., supplier-client relations), standardization was proposed. As all the organizations need to manage the expectations of the clients, the first standard of common importance is a series of ISO 9000 standards (starting one is currently ISO 9001:2015 Quality management systems, requirements; ISO 2015a). While that one commonly requires that the basic organization and principles are set and documented to assure the quality of the products/services towards the clients, usually organizations are facing additional issues. For example, raw materials consumption and pollution and waste generation – environmental impacts – are managed within environmental management systems like ISO 14000 series (starting with ISO 14001:2015 Environmental management systems – requirements with guidance for use; ISO 2015b); next, energy consumption/efficiency is the subject of standard ISO 50001:2016 Energy management (ISO 2016) and occupational health and safety are the subjects of coming standard ISO 45001 Occupational health and safety management systems[1] (ISO 2017),[2] etc. Related to risk management, ISO 31000:2009 Risk management – Principles and Guidelines standard – is available (ISO 2009); however, it should be noted that this one is generic and not process industry risk management specific. Needless to say, all standards build on the well-known Deming's loop principles (Plan-Do-Check-Act steps) are applied for the aspect in question.

The next question related to the management system and details required to be implemented specifically for the management of changes is what exactly is required. Let us briefly examine what is requested in that respect in legislation or guidelines:

- In the EU, the current Seveso III directive (EC 2012) required that a Safety Management System (SMS) is to be implemented (specified in its Annex III). Among seven issues, its fourth states: "(iv) management of change – adoption and implementation of procedures for planning modifications to, or the design of new installations, processes or storage facilities". Not much of a detail and possible additional guidance might be available in national guidelines or textbooks.
- In the United States, legislation requires that the prevention program considers management of change (US 1992; see §68.75 Management

of change) and defines the main items, including "replacement in kind", scope and role of the employees and contractors – in overall 190 words.

- As usual, the devil is in the details. So the practitioners need more detailed guidance on the contents and method to be used in management of changes, e.g., Sanders (2005), section 11, Effectively managing change in the chemical industry (38 pages). That section suggests many practical ideas, approaches and example checklists and forms to be implemented, with complexity depending on the size of organization and operations.
- However, the best reference so far about process safety management systems and specifically on the management of change are CCPS's guidelines on risk based process safety (CCPS 2007, 2008, 2013): the first one in almost 700 pages explains details to be considered that are elaborated in 20 principal elements (no. 13 is management of change), the second one is accompanying and more elaborated on management of (mainly technical) changes, and third one is on the organizational changes only. For the purpose of this chapter, it is important to note that the first one consistently suggests an approach for the implementation of each of the 20 elements following the key principles and more detailed key features. An example for the management of change element and brief explanations is summarized in Table 12.1.

Table 12.1 Summary of key principles, key features for element Management of Change according to CCPS (2007), section 15.2. In addition, key points considered in organizational readiness evaluation chapter are listed

Key principle	Key features	In brief	Key point[a]
Maintain a dependable practice	Establish consistent implementation	Defined MOC policy in specific Standard Operating procedure (SOP)	A. Implement SCM procedure
	Involve competent personnel	Manage basic personnel awareness on MOC procedure and assure specific training to involved personnel	
	Keep MOC practices effective	Assure performance monitoring, KPIs collection, auditing	
Identify potential change situations	Define the scope of the MOC system	Define to which areas, processes, work and types of changes it applies	B. Define scope of SCM

Key principle	Key features	In brief	Key point[a]
	Manage all sources of change	Define replacement in kind (RIK) and what it is not, check for shortcuts	
Evaluate possible impacts	Provide appropriate input information to manage changes	Define accurate and needed inputs to the evaluations	C. Identify possible change impacts
	Apply appropriate technical rigour for the MOC review process	Define single or multiple review process depending on the complexity and competence/expertise needed	
	Ensure that MOC reviewers have adequate expertise and tools	Define and assure the expertise needed by name and position	
Decide whether to allow the change	Authorize changes	Define one or more competent authorizers that are independent of the proposers	D. Decide whether to allow the change
	Ensure that change authorizers address important issues	Authorizers should address the important issues (hazards, risks, measures) and have access to the whole change dossier	
Complete follow-up activities	Update records	Process safety documentation must be duly updated	E. Complete follow-up activities
	Communicate changes to personnel	Employees should understand the changes to be/that were implemented	
	Enact risk control measures	Safety measures (risk controls) might be mandatory to approve and implement the change	
	Maintain MOC records	Define, assure and check the retention period for the changed process safety documentation, incl. change dossier	

Note: [a] – Key points as considered in the elaboration of 54 detailed audit questions in readiness evaluation

Source: Levovnik & Gerbec 2017

To conclude here, for practical implementation in major hazard organizations, the SOP for management of changes should follow at least the key items and features suggested and that shall include suitable performance monitoring, e.g., rate and scope of change proposal, periodic audits of compliance of the practice with the SOP, as well as periodic management reviews by the top management.

One of the authors of this chapter has been also involved as a consultant to various industrial organizations on the subject of major accident hazards and usually organizations formally manage aspects of quality, environment, occupational health and safety on which they added also a (process) safety management system due to statutory requirements (EC 2012). Of those, in addition, one organization managed even six aspects among those listed in the previous paragraph. So what is a bottom line here? Imagine the organization managing at least three or four aspects and facing a continuous flow of improvements/changes proposals in order to respond to market demands. As mentioned previously in the Introduction section, Zwetsloot et al. (2007) found that managers are facing a challenging complexity of changes, and all impacts on the organization's management are hard to predict before they are on the path of implementation. The solution here is obviously in considering the specifics of both technical and organizational dimensions and adding the explicit consideration of various management levels and managed aspects and interactions among them. A graphical illustration of such scoping is presented on Figure 12.2.

Integrated safety management of changes

In order to address the complexity of the changes in process industry, recently (Gerbec 2016) an integrated a method named safety change management (SCM) was proposed. It builds from evaluating separately the technical/technology related impacts from proposed changes, and is followed by the evaluation of organizational impacts over the applicable management levels and aspects. In that respect, it builds on the use of existing guidelines (CCPS 2007, 2008; Sanders 2005; CCPS 2013). This simply means that interactions and mitigation actions are within the analysis mapped into the matrix where on both sides the management levels (see Figure 12.2, for example) are listed. In the cells, if applicable, corresponding impacts identified are documented. Based on that the potentially required mitigations (to compensate for the specific negative impacts of the change in question) are duly planned together

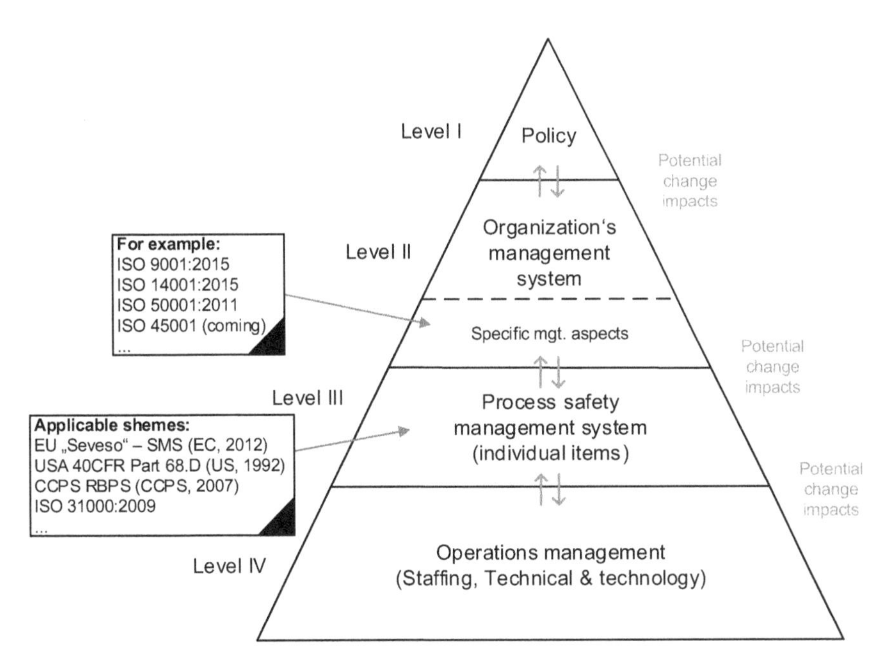

Figure 12.2 Illustration of scoping of the management levels and managed aspects in the organization applied to the potential changes

Source: adapted from Gerbec 2016

with change implementation actions. The method considers also possible temporary changes, the need to document, communicate and train about the specific change – all through the implementation actions that are to be planned, verified and closed by the members of the team. While the method is explained in a detail (Gerbec 2016) and all evaluation forms are available there, the general idea of integrated evaluation of the changes is illustrated on Figure 12.3. It should be understood that details on the method and steps to followed are explained and illustrated in Gerbec (2016) and in Gerbec (2017). Checklists are widely used in the industry both as a reminder and as a findings recording tool in management of change situations, and the same applies for their use in SCM method. The method currently uses seven forms (closely related to the 20 steps presented in Figure 12.3) using in total 273 detailed questions. In order to illustrate the use/data collected within the example for case 1 from Gerbec (2017), checklists for Form IV (Activities mapping and evaluation), Form VI (Evaluation of technical and technology changes), steps 8 and 10 on Figure 12.3, respectively, and Form A, part II, step 1 with interaction matrix and mitigations are presented in Figure 12.4. The reader will note

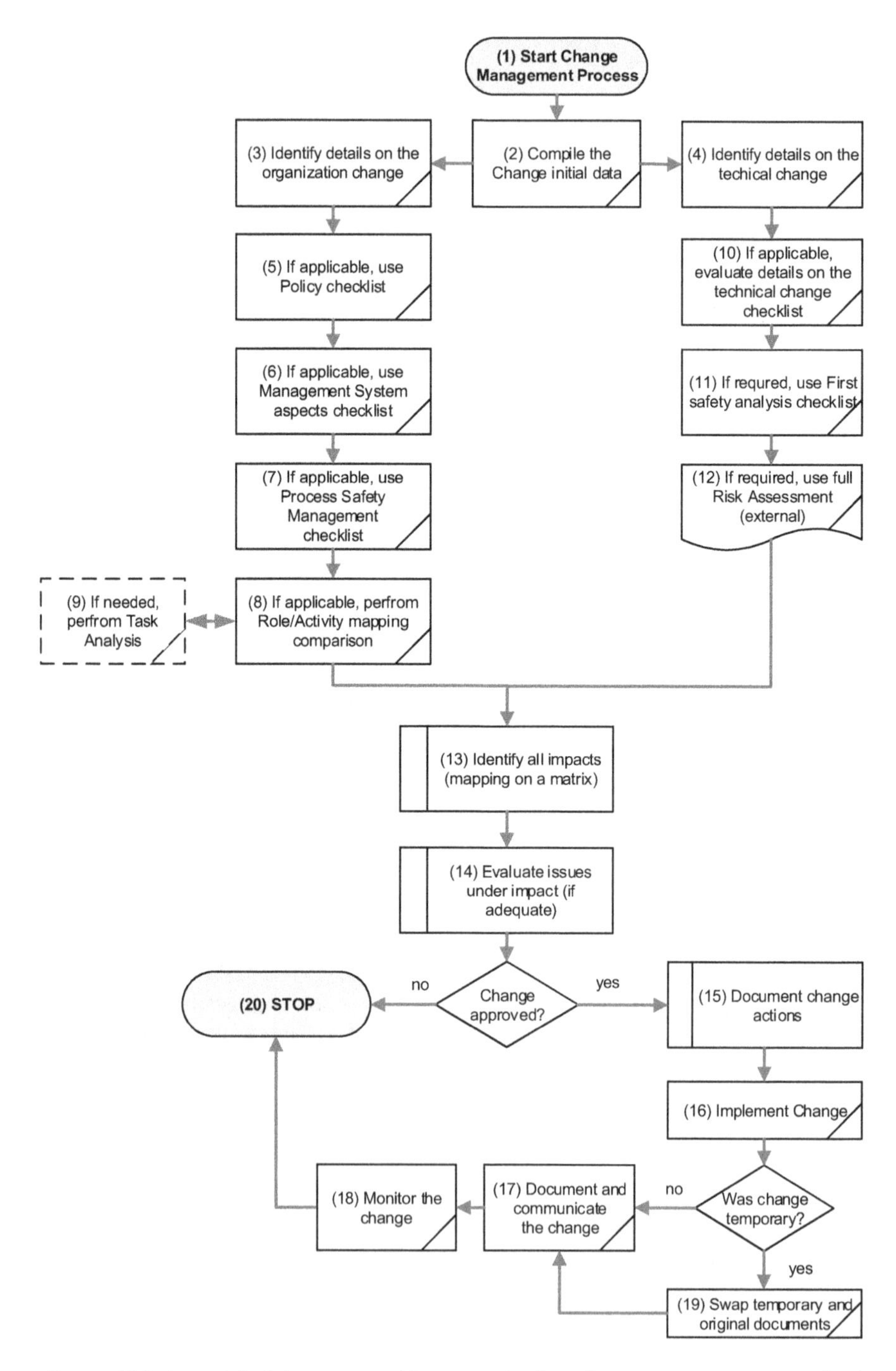

Figure 12.3 A model of the proposed integrated safety change management method

Source: adapted from Gerbec 2016

EXERPT FROM FORM IV: THE ACTIVITIES MAPPING AND EVALUATION

	OLD ROLE ACTIVITIES										NEW ROLE ACTIVITIES								COMPARISON [b]		
				Involves [a]								Any new role issues?									
No.	Appointed activities	Can it be eliminated?	Req. hours per week	Response to deviations	Assuring plant integrity	Assuring plant availability	Managing HSE procedures	Manages essential knowledge and	Subject of change?	Significant risk?	Appointed activities	Req. hours per week	Significant HSE hazards?	High workload, fatigue?	Competence issues?	Communication issues?	Team work issues?	Motivation issues?	Old role control measures that were applied?	Additional measures needed, or any comments?	Will controls be adequate?
1	manual cylinders pressure water washing	yes	40	Low	Low	Low	Low	Low	yes	no	abandoned - the operator is now available for other production	0	no	no	no	no	no	no	(manual washing was labour intensive in stressfull conditions, open space, rain, ...)	none - washing is now done by automated machine in building	yes

EXCERPT FROM: FORM VI – EVALUATION OF THE TECHNICAL & TECHNOLOGY CHANGES

ID, (sub)Component, potential impact(s)		Impacted (requires MOC)?	What is impacted?	Kind of impact?	Possible consequences? [a]	Possible mitigations? [a]	Action by? [a]	Is mitigated impact adequate? [a]
...								
B.	**PROCESSES**							
...								
B.4	Temporary procedural changes	yes	temporary filling procedure needed for the duration of construction works	need to regulate ignition sources	fire and expl. hazard	Works planning	project coordinator	yes
B.5	Changes to raw materials, intermediates or products							
B.6	New raw materials, intermediates or products	yes	Quality: better washing of cylinders	positive	/	/	/	yes
D.	**INSPECTION & MAINTENANCE**							
D.1	Equipment inspection-	yes	new equipment	/	inspections not planned	add to inspections planning	project coordinator	yes
D.2	Procedures & instructions for inspections	yes	insp. & maint. procedure needed	/	procedure missed?	prepare procedure	project coordinator	yes
D.3	Changes to the equipment set-up	yes	chain transporters need repositioning	operations	unable to provide temporary fill-in	organize temporary fill-in	project coordinator	yes
...								

part of examples

FORM A, PART II, step 1 – INTERACTION MATRIX AND MITIGATIONS

		Inputs – please enter Form I to VII Categories IDs in related cells for issues identified				
		I	II	III	IV	VI & VII
	Impacted:	Organization's policy	Organization's management system	Organization's PSMS	Activities and safety critical roles	Technic or technology & safety assessment
1	Organization's policy					
2	Organization's management system					
3	Organization's PSMS					
4	Activities and safety critical roles					B.4, B.6, D.1, D.2, D.3, D.4, D.5, D.7, D.8, D.9, D.10, E1-E2, E4-E6, E9, E11-E13, F2
5	Technic or technology & safety assessment				IV.1	

Used for planning of implementation actions (includes possible mitigations)

Figure 12.4 Illustration of application of Forms IV and VI and Form A (interaction matrix) from Case 1 excerpt

Source: adapted from Gerbec 2017

that, for example, Form VI uses a structured approach in identifying technical changes under impact, and for each topic the potential impact (yes/no), description, consequences, mitigations and appointed actions, as well as final decision on the adequacy, are documented. In the current example, in total 22 impacts were identified (not all details are shown for the sake of brevity) and were mapped to cells IV.5 and VI/VII.4. Based on the detailed impacts identified, the next step was to list 20 summarized impacts and 13 implementation actions (details are available in Gerbec 2017).

Let us conclude here that the proposed SCM method was the subject of testing and validation in three industrial cases of recent changes implemented (ex-post evaluation). The involved plant managers reported that the tested method is more demanding in terms of required resources and time needed; however, on the very positive side, it provided a valuable and very detailed insight into the change implications in a transparent and documented way that allowed following the established standards and aspects by their elements via prepared checklists. In other words, a common operational picture about the proposed change(s) can be established and complex changes can be better managed.

Readiness of management for changes in industry

Having explained the field of management of the technical as well as of the organizational changes, thus introducing the proposed safety change management method in the previous subsection, we might continue to the issue of actual practices used in the major hazard industries. For that purpose, we recently proposed a new method for evaluation of the organizational readiness (maturity) following the Nertney wheel principles (Nertney 1987; Frei et al. 2015) over the activity of safety change management[3] in organizations. The purpose of development of the method was to facilitate the identification of the possible gaps in safety change management in organizations (thus serving as a vehicle to propose opportunities for the improvements), as well as to allow benchmarking among different organizations or their departments, as commonly asked for by the managers.

For that purpose we considered the elaborated specification of the key items and features (e.g., CCPS 2007, section 15.2), as summarized on Table 12.1, interpreted the requirements considering also integration principles suggested by the safety change management method (Gerbec 2016) and prepared a list of key items (Table 12.1, right most column) and a corresponding set of 54 detailed key features and example (candidate) audit type questions (Levovnik & Gerbec 2017). That allows to identify detailed performance issues. Next, the findings have to be mapped to the Nertney wheel stages. Following the Nertney principle, the industrial organizations shall treat their operational systems and their parts as whole: people, procedures, equipment and conducive conditions. The concept builds on three basic elements of any system:

- People (personnel)
- Hardware (process equipment/tools; buildings and grounds)
- Procedures and management controls.

Previously mentioned conducive conditions can be understood by the interfaces among the three elements, where they meet (Nertney 1987, p. 4):

- People – hardware interface match (where hardware is operable by people)
- Procedures – hardware interface match (where hardware is supported by related procedures)
- Procedures – people interface match (where procedures are related to the people).

In other words, procedures must correspond to the actual equipment as it is and must match the needs and abilities of the people. Figure 12.5 (left) represents a graphical representation of the three elements and of the three corresponding interfaces (consisting of its six slices); on the right side it presents six slices elaborated to six activity developmental stages through an operational readiness cycle, e.g., started by the definition of the operation, followed by a number of detailed steps, and likely ending with its release for operation, not to mention monitoring and review (Frei et al. 2015, section 3.1 of Frei et al. 2015). The Nertney wheel principles do not provide us with individual developmental stage criteria, but only with the functionality to be provided at each stage, per each element (subsystem) and per each interface. In order to adopt the Nertney wheel concept for the assessment of the organizational readiness of safety change management, the generalized stage names and definitions have been prepared as presented on Table 12.2.

The readiness progresses from the outer circles towards the central circle ("GO"), following the criteria, that in order to progress to the next circle/stage, the pertaining criteria for each subsystem and each interface for the current stage need to be fulfilled first. It is about a collective state of readiness in accordance with system (activity requirements, Nertney 1987, p. 6).

Having defined the detailed requirements per each stage and its elements, mapping of the prepared 54 detailed audit questions was performed

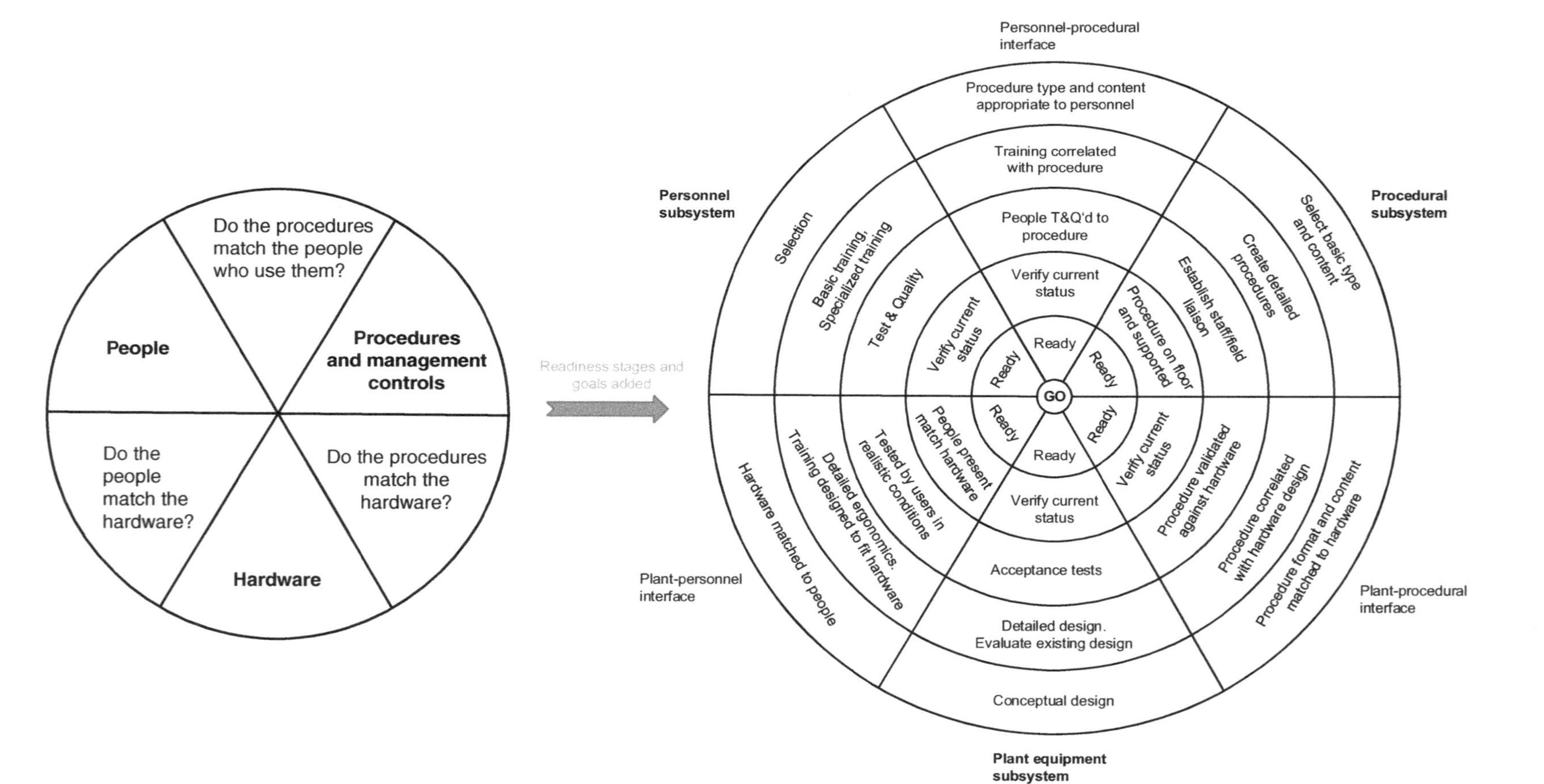

Figure 12.5 Graphical presentation of the Nertney wheel elements and interfaces and six developmental stages

Source: adapted from the Noordwijk Risk Initiative Foundation (Nertney 1987; Frei et al. 2015)

Table 12.2 Overview of the generalized definitions of Nertney's readiness stages as interpreted by Nertney (1987) and Frei et al. (2015)

Stage	Stage name	Definition
1	Basic principles defined	Basic procedural, personal and equipment requirements, including their interfaces, are defined for a specific management activity/ operational system
2	Detailed principles defined	In addition to basic, also detailed requirements are clearly defined for procedural, personnel and equipment parts, including their interfaces. This includes compliance with best practices/ recommendations.
3	Principles subject of testing and qualifications	In addition to detailed principles defined, performance testing and qualifications are implemented and performed, applying to all parts and their interfaces
4	Principles verified, including interfaces	In addition to testing, internal certification processes are defined and performed related to all three parts & interfaces within the managed activity/operational system
5	Ready	Specific subsystem/interface ready
6	GO!	Plant/equipment, people, procedures are congruent – GO!

as criteria on them. The detailed mapping will not be provided here for the sake of brevity (for full reference please refer to Levovnik & Gerbec 2017). The analysis shall consist of the following actions:

1 Auditor(s) perform audit of the organization considering the presented evidences, interviewed auditees and document the audit questions, findings to the questions (yes/no), comments and potential improvement suggestions.
2 Findings are mapped per readiness (maturity) stages and subsystems & interfaces according to the criteria set. The criteria, comparison to the findings and stages reached are presented in a form of the table (the wheel representation would not be suitable).
3 Visualization and interpretation of the findings is carried out as the overall readiness stage reached by the organization (benchmarking) as well as in profiling the readiness stage reached by each subsystem and interface.

For the practical application, a specific spreadsheet tool was prepared in order to automate the analysis and interpretation of the readiness stages. Using the tool, we performed testing and validation of the method in nine industrial organizations subject to major accident hazards. Summarized data on the anonymous test organizations is provided in Table 12.3.

The detailed explanation of the results obtained and their interpretation will be reported in a coming publication; however, here we will present only most important findings from the testing. The profile of the readiness stage reached by the organizations by subsystems and interfaces is presented on Figure 12.6.

Table 12.3 Summary of info on nine anonymous industrial organization involved in testing the proposed method

Organization	NACE[a] classification	Number of employees[b]	Classification[c]
A	C 20.2 Manufacture of pesticides and other agrochemical products C 20.1 Manufacture of basic chemicals, fertilizers and nitrogen compounds, plastics and synthetic rubber in primary forms	Medium	Upper tier
B	C 20.13 Manufacture of other inorganic basic chemicals	Medium	Upper tier
C	B 06.10 Extraction of crude petroleum.	Large[d]	Upper tier[e]
D	D 35.11 Production of electricity	Medium	Lower tier
E	C 20.16 Manufacture of plastics in primary forms	Medium	Upper tier
F	C 20.11 Manufacture of industrial gases	Medium	Upper tier[e]
G	G 47.30 Retail sale of automotive fuel in specialized stores	Large	Upper tier[e]
H	D 35.30 Steam and air conditioning supply	Large	Upper + Lower tier[e]
I	C 20.30 Manufacture of paints, varnishes and similar coatings, printing ink and mastics	Large	Upper tier[e]

Notes:

a – According to Eurostat (2008).

b – Micro: <10; Small: 10–49; Medium: 50–249; Large: >250 employees (Eurostat 2008).

c – Lower or upper tier category according to the national legislations implementing EC (2012).

d – More than 1,000 employees on the related production site.

e – Operations on more than one site (different categories according to note [c] may apply).

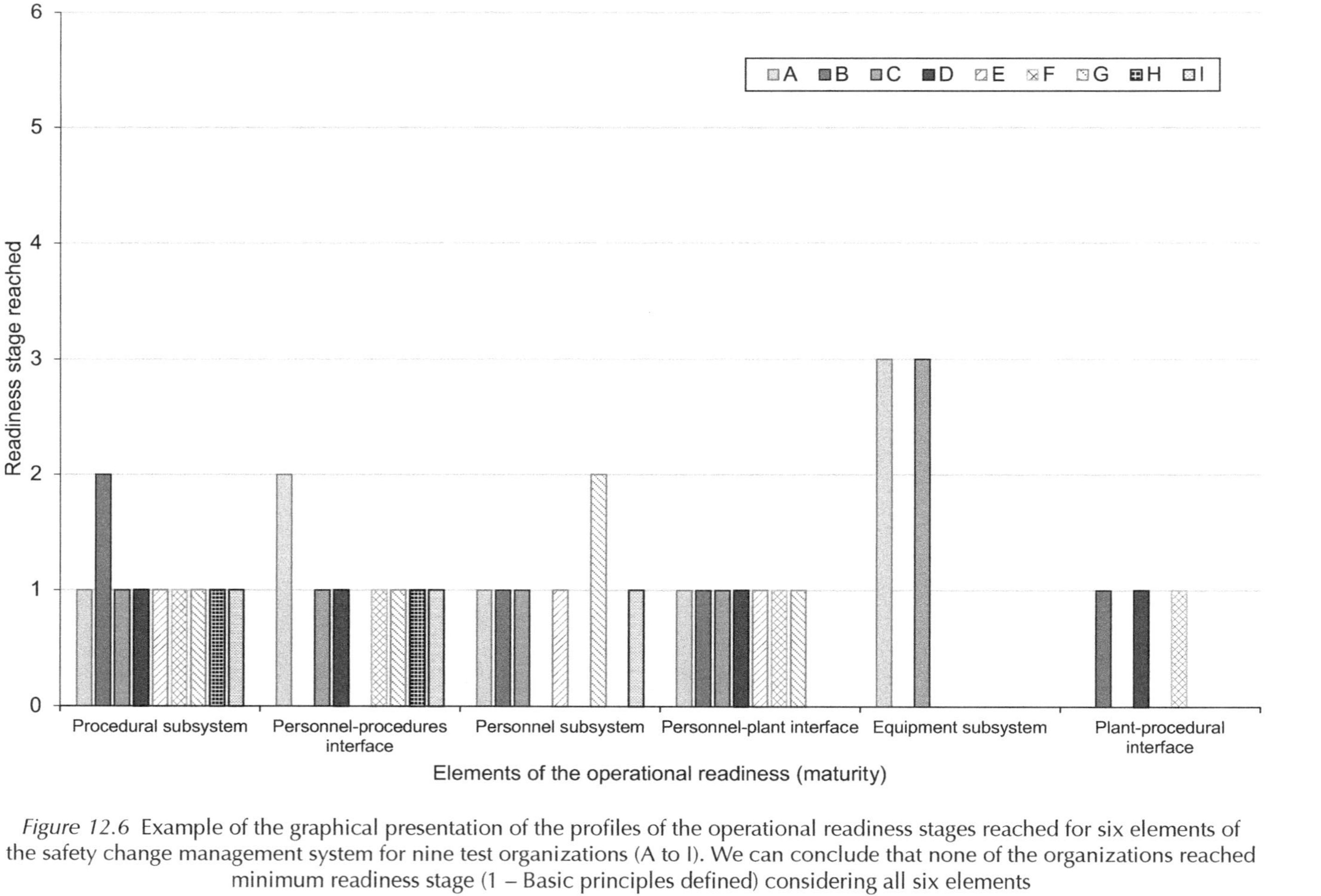

Figure 12.6 Example of the graphical presentation of the profiles of the operational readiness stages reached for six elements of the safety change management system for nine test organizations (A to I). We can conclude that none of the organizations reached minimum readiness stage (1 – Basic principles defined) considering all six elements

It can be observed that individual organizations obtained different readiness stages per each of six the elements. For example, organization A reached stage 1 (readiness of basic principles defined) in procedural and personal subsystems and personnel-plant interface, stage 2 (readiness of detailed principles defined) in personnel-procedure interface, stage 3 (principles subject of testing and qualifications) in equipment subsystem. However, it did not reach even the stage 1 in plant-procedural interface; according to Nertney wheel principles; thus it is at developmental stage 0 (a kind of being "outside" of the circles as shown on Figure 12.5). Similar conclusions can be reached for the rest of the test organizations, namely, none of them reached stage 1. This comes as a kind of surprise. The list of key items required and the Nertney wheel principles are demanding in that respect. For example, organization A failed to reach stage 1 due to negative findings related to key feature in plant-procedural interface, specifically at the audit question B.1 "Is the scope of technical changes defined in detail (extensive lists, referencing applicable guidelines and practices)?" The negative finding is based on the reported fact that simple checklists are used, but the details to be examined, assurance on their completeness and referencing on them were not available. Similar detailed explanations are available also for higher stages and findings per all organizations. In the case of organization A, it appears that the only deficiency found at stage 1 could be quickly overcome by adding transparency and traceability to the existing procedure.

Considering all the organizations and all the readiness stages, the following can be summarized: the overall understanding of safety change management among the auditees was good. In most cases, organizations established some sort of informal or partly defined management of changes (safety change management); generally, procedures and criteria were not clearly defined and the term "replacement in kind" (as the opposite of the change) is not commonly defined. Managers and experts are not specifically trained on their roles in safety change management, and analytical methods and tools are undefined and remain at their personal selection. Big issues in stage 3 are absence of use of any performance indicators, no planning or performing internal audits and management reviews. On the other side, the general change goals are pre-defined and evaluated for large parts of the organizations. To conclude, the auditees were, after the interviews, interested in the results and in the readiness stages reached. As a rule, they commented that the benchmarking is interesting to them, considering the confidentiality. As none of them reached the basic readiness stage, the question at hand was what exactly is missing and how to improve things.

Conclusions

The management of changes in major hazard establishments is on one side part of mandatory safety management systems, and current understanding usually considers evaluation and planning of proposed technical changes about potential implications on process safety. On the other side, narrow consideration of only technical changes is deemed insufficient due to the complexity and scope of the changes – thus a comprehensive evaluation, decision and implementation is needed to consider interactions among the organizational and technical dimensions (e.g., using the method proposed in Gerbec 2016). The formal requirement for the management of changes has been in place here for more than 20 years, and it was worth it to take a look at how well some industrial organizations are doing in preparing for and practicing its use. The evaluation of the overall operational readiness (maturity) stages according to the Nertney wheel principles on the key features of the management of change in a form of structured audit interviews on nine test organizations provided surprising results: rarely the organizations consider organizational changes for safety implications, never for integrated evaluation of both technical and organizational dimensions, and even more fundamentally, the internal procedures for evaluation are usually not duly documented, not to mention there are no formal training of the evaluators, approvers, applicable tolerability criteria, monitoring performance (e.g., KPIs), audits and reviews, etc. In other words, leaders in the industry seem to recognize the importance of the changes, but they still rely on personal competence. The accumulated experience is not documented in order that the knowledge obtained can be added to the organization's repository of knowledge. In that respect, we recall the late Trevor Kletz's proverb: "Organizations have no memory, only people have memory", which is to be understood as a guide to "Conserve the knowledge of the system and the logic behind the decisions" (Frei et al. 2015, p. 13), by defining, using and continuously improving the procedures.

Thus, it seems that we need to start at the beginning and educate graduate students in technical disciplines as well as to communicate and train the current set of leaders in the industrial organizations, for example, through the industrial associations. Last, but not least, the same likely applies also for the regulators that shall effectively supervise the industrial organizations in question.

Notes

1 ISO 45001 standard aims to replace the current British OHSAS 18001 standard by 2018.
2 Related to food risk management is the HACCP (Hazard analysis and critical control points) requirement or coming standard ISO 22000 Food safety management systems. Requirements for any organization in the food chain are to be mentioned.
3 The term "safety change management" as used in this chapter relates to the common management of changes in organizations for the aspect of process safety.

References

Balfe, N, Leva, MC, Ciarapica-Alunni, C & O'Mahoney, S 2016, 'Total project planning: Integration of task analysis, safety analysis and optimisation techniques', *Safety Science*, vol. 100, pp. 216–24, http://dx.doi.org/10.1016/j.ssci.2016.10.014

Cameron, E & Green, M 2012, *Making sense of change management: A complete guide to the models, tools and techniques of organizational change*, Kogan Page, London, Philadelphia & New Delhi.

CCPS 2007, *Guidelines for risk based process safety*, Center for Chemical Process Safety, Wiley, Hoboken, NJ. ISBN 978-0-470-16569-0.

CCPS 2008, *Guidelines for the management of change for process safety*. Center for Chemical Process Safety, Wiley, Hoboken, NJ. ISBN 978-0-470-04309-7.

CCPS 2013, *Guidelines for managing process safety risks during organizational change*. Center for Chemical Process Safety, Wiley, Hoboken, NJ. ISBN 978-1-118-37909-7.

CSChE 2004, *Managing the health and safety impacts of organizational change*, Canadian Society for Chemical Engineering, Ontario, viewed 13 January 2016, www.cheminst.ca/sites/default/files/pdfs/Connect/PMS/Managing the Health and Safety Impacts of Organizational Change.pdf

EC 2012, Directive 2012/18/EU of the European Parliament and of the Council of 4 July 2012 on the control of major-accident hazards involving dangerous substances, amending and subsequently repealing Council Directive 96/82/EC; viewed 13 January 2016, http://eur-lex.europa.eu/legal-content/EN/TXT/?uri=CELEX:32012L0018

Endsley, M, Bolte, B & Jones, D 2003, *Designing for situation awareness: An approach to user-centered design*, Taylor Francis Group, Boca Raton.

Eurostat 2008, 'Eurostat methodologies and working papers', *NACE Rev. 2: Statistical classification of economic activates in the European Community*. ISBN 978-92-79-04741-1, viewed 22 May 2017, http://ec.europa.eu/eurostat/documents/3859598/5902521/KS-RA-07-015-EN.PDF

Frei, R, Garforth, A, Kingston, J & Pegram, J 2015, *Using operational readiness to improve the management of risk, volume 1: Concepts*, The Noordwijk Risk Initiative Foundation. ISBN 978-90-77284-12-4, viewed 8 May 2017, www.nri.eu.com/WHITE%20PAPER%202.1.pdf

Gerbec, M 2016, 'Safety change management: A new method for integrated management of organizational and technical changes', *Safety Science*, vol. 100, pp. 225–34, http://dx.doi.org/10.1016/j.ssci.2016.07.006

Gerbec, M 2017, 'Management of technical and organizational changes', in M Čepin & R. Briš (eds.), *Safety and reliability: Theory and applications, 27th European safety and reliability conference, ESREL 2017, June 18th–22th, 2017, Portorož, Slovenia*, CRC Press Taylor & Francis, Boca Raton, 1595–1602, https://doi.org/10.1201/9781315210469-200

Hayes, J 2010, *The theory and practice of change management*, 3rd edn, Palgrave Macmillan, Basingstoke & New York.

HSE 2016, UK HSE website guidance *"plant modification/change procedures"*, viewed 20 April 2017, www.hse.gov.uk/comah/sragtech/techmeasplantmod.htm

ISO 2009, ISO Standard 31000:2009(en), Risk management: Principles and guidelines, viewed 15 January 2016, www.iso.org/obp/ui/#iso:std:iso:31000:ed-1:v1:en

ISO 2015a, ISO Standard 9001:2015(en), Quality management systems: Requirements. ISO, viewed 15 January 2016, www.iso.org/obp/ui/#iso:std:iso:9001:ed-5:v1:en

ISO 2015b, ISO 14001:2015(en), Environmental management systems: Requirements with guidance for use, viewed 13 June 2017, www.iso.org/obp/ui/#iso:std:iso:14001:ed-3:v1:en

ISO 2016, ISO 50001:2016 Energy management: Requirements with guidance for use, viewed 13 June 2017, www.iso.org/obp/ui/#iso:std:iso:50001:ed-1:v1:en

ISO 2017, Draft ISO/DIS 45001.2 Occupational health and safety management systems: Requirements with guidance for use, viewed 13 June 2017, www.iso.org/obp/ui/#iso:std:iso:45001:dis:ed-1:v2:en

Levovnik, D & Gerbec, M 2017, 'Auditing operational readiness of management of change', in M. Čepin & R. Briš (eds.), *Safety and reliability: Theory and applications, 27th European safety and reliability conference, ESREL 2017, June 18th–22th, 2017, Portorož, Slovenia*, CRC Press Taylor & Francis, Boca Raton, 1659–1666, https://doi.org/10.1201/9781315210469-209

Nertney, RJ 1987, *Process operational readiness and operational readiness follow-on*, System Safety Development Center, SSDC-39, DOE-76-45/39, viewed 8 May 2017, www.nri.eu.com/SSDC39.pdf

Paton, RA & McCalman, J 2008, *Change management: A guide to effective implementation*, 3rd edn., Sage Publications, London.

Sanders, RE 2005, *Chemical process safety learning from case histories*, 3rd edn., Elsevier Butterworth & Heinemann; Amsterdam, Boston. ISBN 0-7506-7749-X.
Seppanen, H, Mäkelä, J, Luokkala, P & Virrantaus, K 2013, 'Developing shared situational awareness for emergency management', *Safety Science*, vol. 55, pp. 1–9.
US 1992, 40 CFR part 68, Subpart D: Program 3 prevention program, viewed 13 January 2016, www.ecfr.gov/cgi-bin/text-idx?SID=243639l8a1da6e6c82f419fa327e6979&mc=true&node=sp40.16.68.d&rgn=div6
Zwetsloot, GIJM, Gort, J, Steijger, N & Moonen, C 2007, 'Management of change: Lessons learned from staff reductions in the chemical process industry', *Safety Science*, vol. 45, pp. 769–89.

Risk and productivity

The way forward

David Levovnik, Marko Gerbec and Maria Chiara Leva

Introduction

The first chapter has elaborated on the evolution of safe work systems, occupational safety and process safety, legislation, standards and elements within the Total Safety Management (TSM) framework. The following chapters presented related methods and cases applied to the TSM that were mostly developed within the TOSCA project and illustrated how TSM can be implemented in real organisations.

However, the work on the development and implementation of the TSM framework is far from being finished. Results obtained from the case studies, feedback/questions from the end users, as well as difficulties encountered and reported by the researchers strongly suggest that a number of topics require further work. This chapter aims to explain some further research topics of importance to the successful implementation of the TSM framework:

- *Business value from the safety programs:* there is an issue about how managers perceive the impact of safety management systems and related efforts to the long term business performance of organisations. Safety programs do not generate any direct revenues, so is it about doing the mandatory statutory duty[1] or is it indeed about effective safety programs helping to reduce operational and maintenance costs and costs of accidents and improving the productivity?[2] Careful consideration should be given to the adverse effects of accidents, to how safety is intertwined with other business functions and to how safety correlates to business performance. To conclude, it is expected that extending the risk assessment results beyond the conventional "counting of the dead, the injured and the demolished assets" to the prognosis of potential organisational performance can increase the visibility of safety programs and the required resources in the eyes of the managers.

- *Risk management as a basis for decision-making:* there is an operational issue how to impose a risk based framework on the current decision-making processes of organisations. In this topic, we discuss how the management domain understands safety decisions and how risk management can be integrated in an efficient manner. We also explain why it is important that safety becomes part of a business strategy and how the CEO is in a unique position to facilitate the implementation of safety management into all aspects of the organisation.
- *The value of Safety Performance integrated in business performance:* The importance of implementation of the ISO 31000 risk management standard and the integration of risk management process in the organization's business strategy shall be given particular attention in this chapter. The use of decision analytics and cost benefit analysis in risk based decision-making is also discussed.

How to create business value through safety programs?

Accidents and their adverse effects

Irrespective of the type of industrial organisation, accidents are highly undesired events that can cause damage to life, property and environment. Accidents can have a severe impact that can be seen either right after the event or after a long period of time (Kim et al. 2012). It is important to stress that, in addition to the time elements (i.e., short term and long term impacts), we must also consider a number of entities that are affected. Damage can affect individuals (individual perspective), the wider society (social perspective) or the organisation itself (organisation's perspective). The consequences that are sustained by the society, in the event of an accident, are certainly different from the consequences that are sustained (only) by the organisation and should not be confused when assessing them (Hopkins 1999). Some of them can easily be linked directly to an accident (direct causes) while others cannot (indirect causes). From the organisation's point, accidents can seriously interrupt operational processes, affect the quantity and quality of products and generate a financial loss. Accidents can be the cause of missed delivery times and delays that can lead not only to financial losses, but also to a negative impact on the client's perception of the organisation. From the employee's point of view, accidents can negatively influence their motivation and productivity. At worst, this may lead to the most capable employees leaving the organisation. In such cases, a wide array of unpredicted yet substantial losses may incur (Argilés-Bosch et al. 2014). Because organisations are in the best position to influence the occurrence of accidents, this part will focus only on organisations' view on this matter. When considering the impact of accidents, we must be aware that damage sustained in a case of critical events can be fatal for the future of businesses even if this impact is not immediately evident (Kim et al. 2012). Therefore, safety programs can have important financial effects on the organisation and can even help to create business value.

How safety is connected to other business functions

The relationship between safety practices and the wider organisational context is of utmost importance if we want to understand the role that safety can play in the organisation. Usually, safety research is focused only on the stakeholder's perspective (e.g. workers, government, etc.). Consequently, a wider organisational view may be neglected, even though safety is inextricably intertwined with almost every business function in the organisation. It would appear that safety is also closely related to other organisational goals, on which it may have either a positive or a negative effect. Among business functions, operations management is perhaps the one that is mostly in line with safety. Accidents most likely occur in a production facility where the production system and the operations workers sustain some type of safety relation. If safety conditions are inadequate, human operators may suffer from work related injuries or illnesses even without reporting a formal accident (Veltri et al. 2013).

When we are assessing the relationship between safety and other business functions, we also have to take into account their outcomes. A study performed by Veltri et al. (2013) shows that organisations can improve both safety and operational outcomes at the same time. When workers are able to think about safety and integrate safety practices in a way that is aligned with operational goals, it is then also possible for safety to become connected with business operations. Nevertheless, there are some prerequisites that have to be met. Responsibility for safety has to be shared by all employees and should not be shifted only to safety practitioners. It is important that operational managers, who are responsible for ongoing work, are also accountable for employees' safety. At the same time, it is important that safety practitioners are aware that operational managers may not have the safety relevant knowledge that would allow them to manage safety challenges. Operational managers may have everyday control over safety at the workplace; however, safety practitioners must provide consultation and specific risk mitigation advice to ensure the safety of employees and to increase the operating outcomes at the same time. When this is achieved, it is less necessary for the organisation to make trade-offs between safety and productivity. This notion is important because in practice the trade-offs are usually made in favour of completing the work over safety (Veltri et al. 2013).

Safety programs and business performance

Besides the wide range of damage or loss (usually measured in millions or even billions of euros), which the organisation can avoid if accidents are prevented, there is another group of positive effects that total safety can bring. Positive impacts that safety programs can have on the business performance of the organisation include productivity improvements or capabilities to prevent negative business results that can be caused by deviations from the foreseen business plan. In case of an accident, only a few organisations can withstand disturbance and follow the business strategy in an exact way that was planned in the first place (Argilés-Bosch et al. 2014).

Productivity improvements

One way to enhance business performance is by influencing employees' performance through safety interventions. Health and safety measures can increase productivity that can consequently lead to positive financial impacts. It has been shown that employees' motivation positively affects their performance. This consequently indicates that health and safety should get appropriate attention if an organisation wants to increase employees' productivity. Also, another important element is ergonomics improvements. In most industries, inadequate ergonomics is the main cause of safety hazards for employees. Ergonomics can eliminate or mitigate safety problems at work and even improve operator performance. Another important element that is frequently overlooked is the site layout planning. It is vital to ensure safe and efficient operations, minimise travel time, reduce unnecessary material handling or prevent potential obstructions (Ayessaki & Smallwood 2017).

There is also the connection between the implementation of safety standards (e.g., OHSAS 18001, forthcoming ISO 45001) and employee productivity that has to be considered. Safety standards can be seen as a strategic tool that could enable organisations to develop several competitive advantages. In a business environment, this is especially important because, in a competitive environment, anything that can give a competitive business advantage is important to the organisation. The impact of safety standard on productivity can be twofold. On one hand, improvements in employee productivity may arise from better health and safety risk management that reduces the number of accidents and the productivity loss related to workplace accidents (e.g. loss of human capital when a skilled employee is injured). On the other hand, the implementation of safety standards and accumulated experience in health and safety management may decrease the probability of occurrence of any disturbances in operational processes. Consequently, this has a positive impact on productivity. There is growing support for the argument that safety knowledge and experience have potential benefits that are much greater and significantly exceed safety outcomes. Safety systems can have an important strategic role. Safety standards can be viewed as an important strategic component that improves workplace safety and, at the same time, supports development of competitive advantage and consolidates operations (Abad, Lafuente & Vilajosana 2013).

Deviation from the business plan

Another way in which safety can influence business performance is the negative financial impact that accidents may have on the organisation in the long term future. Accidents are unexpected events that may not have a significant short term financial impact on the organisation but undoubtedly could cause serious long term damage. They can introduce disruptions to daily operations of the organisation and ultimately result in deviations from the strategic activities that were planned before the accident took place. Full financial consequences of accidents are usually not evident for some time. This is because when an accident occurs, the workforce must redirect their focus to deal with the consequences of this event and at the same time ensure that organisational operations continue to be normal on a daily basis. Argilés-Bosch et al. (2014) believe that daily activities are not the ones that are most compromised. Quite different are the strategic planning activities. They are usually postponed in order to ensure continuity of operations. In many cases, different strategic tasks (e.g. product development, quality assurance, process design, resource planning etc.) usually receive the least attention when such unwanted events occur. In other words, the focus is redirected from strategic to other urgent activities that require imminent attention. As strategic planning is an important business activity that cannot be simply postponed, this can have serious negative effects on later financial performance. As Argilés-Bosch et al. (2014) have shown in their study, negative effects on an organization's financial performance tend to be most significant one year after the undesired event. However, we may anticipate that the level of disturbance sustained by the organisation in the case of accident can depend on many factors such as the size of organisation, the specific nature of the industry and also the management response system (Argilés-Bosch et al. 2014). In addition, there are also external factors (the social, economic and political environment) that have important effects on the financial performance of an organisation (Forteza, Carretero-Gómez & Sesé 2017).

In practice, it is very hard for managers to accurately quantify the financial losses of workplace accidents and compare them to the gains that can be made from sound safety investments. The problem may partially lie in current accounting techniques that make the assessment of actual loss and the prediction of the potential returns on safety investments especially difficult. It may appear that better suited accounting techniques have to be developed and employed to assessing the financial consequences of undesired events. Accident prevention is an important part of the social responsibility of successful, modern organisations and can be used in developing relationships with different stakeholders. Especially for the investors, this information can provide important insight into an organisation's profitability (Argilés-Bosch et al. 2014).

Organisation and its reputation

Reputation is usually an elusive concept. However, its importance when it comes to business value cannot be overemphasised. In the past, a majority of an organisation's value consisted of tangible assets. Nowadays, roughly about three-quarters of an organisation's value is intangible. An organisation's most valuable assets become its brand, good name and of course its reputation (Brigham & Linssen 2010).

As the reputation becomes increasingly important, organisations cannot afford to maintain inappropriate safety conditions, let alone be responsible for large accidents. In the process industry, a major accident can have a significant negative impact on the reputation of the organisation. There are different stakeholders (e.g. clients, suppliers, shareholders, etc.) that can be affected. The organisation can lose an important customer or can have problems with the supplier. In addition, the decline in share value can be seen as a negative signal for the potential investors. In the worst case, it can seriously affect the stability of the organisation, it can impair its growth and it can leave the organisation vulnerable to hostile acquisition. Therefore, the future career of executive managers in an organisation may also be endangered. A recent study (Makino 2016) that examined the response of stock market after an industrial accident has found that the returns on stock were notably lower than expected. If the organisation is forced to file for bankruptcy, shareholders are facing an even bleaker scenario. In other words, shareholders can also sustain serious financial losses.

Risk management: a necessary evil or a key to better decision-making?

Understanding how business decisions are actually made

Making decisions is at the heart of every manager's work. When we look across the organisation, managers are always involved in some type of work that includes decision-making (Brousseau et al. 2006). Therefore, this is the place to start this section. In general, a decision can be seen as a choice that is made between different alternatives. However, making a choice is only one very important part of the decision-making. If we look closely, we could see that decision-making is really a process in which people must identify a problem, search for opportunities and resolve it. Making a decision involves different activities before and after the choice is made (Daft 2008). Managers are engaged in a spectrum of activities including information gathering, reviewing data, developing different ideas, evaluating possible alternatives, implementing the decision that has been made or following its results (Brousseau et al. 2006).

In practice, not every decision is the same. There are a number of different decisions that have to be made daily in every business organisation. We can divide them into four types of business decisions with regard to the level of control over the decision that can be achieved and how success is measured. According to Rosenzweig (2013), decisions can fall into the following categories:

- making judgements and choices
- influencing outcomes
- placing competitive bets
- managing for strategic success

Throughout the different daily tasks, managers have to make many diverse decisions that usually fall into these four groups. In order to make better decisions, they have to understand in which group a particular decision belongs. Managers have to be able to identify properties of the decision they are currently considering. In addition to this, they should be able to use an appropriate decision-making approach that will result in the best possible decision (Rosenzweig 2013). We would like to emphasise that understanding the decision-making process was not always the same through the last couple of decades. According to Soll, Milkman and Payne (2015), there are three different schools of thought that currently have a primacy in this area.

The first is the so-called decision analysis approach which assumes that decision-making is a rational process that can be done systematically, even when conditions of uncertainty are present. Tools such are Decision trees and Bayesian statistics can be used to facilitate the decision-making process. Although the process is rational and consistent, it does not consider the time constraints and the fact that relevant data are not always available. In addition, there is also a question of human cognition. This approach can be used for important long term decisions where reliable data are available.

The second approach, that is fundamentally quite different, is heuristics decision-making (also known as rule of thumb). This approach has often proved to be very effective in practice since it is simple and does not require much additional information. However, we cannot predict whether this approach will actually work in a specific scenario. It is useful when situations are predictable but also where uncertainty is high and we cannot rely on acquired data.

The third is the heuristics and biases approach which is currently the most widely acknowledged and studied. It researches the ways in which people's decisions differ from the mandates of the rational model. Decision-making that is based on heuristics can result in either good or bad judgement. This approach does not prescribe exact instructions for making a good decision; however, it highlights the most common errors, so they can be avoided when making decisions.

In a business world, managers have to deal with many diverse situations that require different types of decisions. Therefore, a combination of different approaches that includes rational decision analysis, bias avoidance and heuristics should be used to make smart decisions (Soll, Milkman and Payne 2015).

Decision-making in the light of risk management

Making important decisions also involves dealing with different types of risks. Risk management is indispensable if the managers want to ensure a stable and successful future for the organisation. Risk managers usually focus on strategic, operational, legal, compliance and financial risks. According to the literature, there are several effective ways to assess and manage an organisation's risks ('How to live with risks' 2015):

- A right balance should be struck between risk and potential reward. In the practice topic of risk management, it is usually associated with risk prevention. For managers, however, low risk also usually results in low returns. Risk managers should, therefore, help the organisation to determine tolerable risks and support decision-making.
- The focus should be on decisions. Risk management is often seen as a task that has no clear effect and results in additional paperwork without any risk reduction effect. Risk managers should employ more innovative approaches and support employees at risk assessment. There can be also improved apprehension of risk in decision-making.
- Include other employees in making risk decisions. People make decisions and risk managers cannot be available at all times for every decision. Therefore, the organisation should train their employees to be able to assess and use adequate risk levels when making decisions.

With newly emerging risks that can threaten organisations, sustainability risk management is gaining significance importance. The main reason behind it is the changing perception of an organisation's business value. Nowadays, a large part of the organisation's value is intangible; hence, different risk assessment approaches should be considered. It is estimated that various non-financial elements may present risks that can go up to more than 20% of an organisation's total market value. In most cases, the main sustainability risk results from either environmental risk, social and ethical risks (in which health and safety present the most important issue), or economic and socio-economic risks. Management should identify risks that present the most important threats to the organisation´s value and utilise adequate measures to effectively manage these threats (Spedding & Rose 2007).

The key challenge here is not so much how to incorporate risk management into decision-making that can have implications for the future of the organisation; there are already some examples of good practice available. Successful organisations perceive every decision as if it could bear some risk. In such organisations, the risk is treated very meticulously and risk management is involved in the work of each manager in the organisation ('How to live with risks' 2015). The main challenge, therefore, is how to incorporate safety into risk management that already exists in the organisation. It is claimed that, in order to implement safety perspective in the organisation's decision-making, it is necessary to ensure that safety becomes an important part of the organisation's strategy.

Safety must become part of business strategy

As noted earlier, it is crucial that safety becomes incorporated into the strategy of the organisation. The influence that the right strategy can have on the success of the organisation is clear and managers are strongly invested in developing the winning strategy. The strategy is similar to a "master plan" in which organisations declare planning to reach their primary mission and outlined objectives. Through strategy, the organisation tries to enhance competitive advantage and reduce competitive disadvantage. At the top there is a corporate strategy that sets out the organisation's overall course regarding its perspective on the development and management of businesses. Next follows the business strategy, which is at the level of the business unit. Its focus is how to enhance the competitive position of organisations. A carefully planned strategy can be crucial to an organisation's long term existence; hence, it should receive special attention from managers. However, it does not end there. Strategy implementation, evaluation, control and learning processes are also equally important (Wheelen & Hunger 2011).

It is also clear that we can significantly improve organisational performance if we achieve to connect measurement of performance and its business strategy (Yuliansyah, Gurd & Mohamed 2017). Nevertheless, most organisations fail to close a gap between strategy development and its successful implementation. One way to achieve this is to use a balanced scorecard approach as proposed by Kaplan and Norton (2007). This approach broadens traditional measures (financial measures) with performance measures that include also the perspective of customers, the organisation's learning and growth, and its internal business perspective. It allows the organisation to simultaneously control other non-financial results that are also relevant to the business (Kaplan & Norton 2007). However, the particular methods of how to successfully include safety into an organisation's strategy and how to incorporate them into an organisation's goals still remain a challenge. A possible application of safety related goals into the balanced scorecard has been, to some extent, addressed by some authors (Mearns & Håvold 2003; Tappura et al. 2015).

The CEO perspective

In practice, the organisation has to pursue different goals that can be sometimes in conflict among themselves. As James Reason noted quite some time ago, we must realise that organisations cannot only be concerned about safe operations. That won't keep them in business. However, we must also be aware that there are not many organisations that could survive a major accident (Reason 1998). This means that when defining an organisation's goals, a fine balance has to be struck. Here, the chief executive officers (CEO) are in a unique position. Only the CEO has the power to connect the inside view (the organisation) and the outside view (the economy, society, markets, technology and customers). The majority of employees in the organisation have an extremely narrow view. They are focused on their specific field of operations and in a specific direction (in most cases employees are inward focused). Therefore, the CEO must use this advantage and integrate both outside and inside perspectives. This is his job because without one there is no other, so they are both equally important.

He is accountable for the organisation's performance and results. The CEO must achieve not only his goals but also fulfil diverse and conflicting expectations of external stakeholders. There are four basic tasks that should be highlighted (Lafley 2009):

- It is important to define what external stakeholders matter the most and what results count the most.
- Decide where to compete. What are the organisation's strengths that give them a competitive advantage, and what are their weaknesses that have to be overcome if an organisation wants to succeed? Adopting the right strategy is a difficult decision; however, because of his broad view, the CEO has the best position.
- Find the right balance between present and future activities. This ability does not come from plain facts, but mainly from longtime experience and good judgement. The CEO must connect both internal and external interests.
- Values and standards (norms) are also important. If values determine the identity of the organisation (true values can lead organisation in the right direction), standards are all about expectations (they can be used as a guide for decisions).

As we can see, CEOs are in the unique position that gives them not only good inside but also outside perspective; they are the ones that are most suitable to decide which business path the organisation should choose. They can define the present and future goals and shape the values and the standards of the organisation. Therefore, they are also in the unique position to take actions that are necessary for implementing safety in all aspects of the organisation, starting at the top with the strategy itself (Lafley 2009).

The value of Safety Performance integrated in business performance

The ISO 31000 Standard (ISO 2009) can be applied throughout the life of an organisation for a wide range of activities, including strategies and decisions, operations, processes, functions, projects, products, services and assets. This standard can be applied to any type of risk, whatever its nature regardless of their positive or negative consequences. The standard lists a set of good reasons for implementing a risk management process as part of a company business strategy. The main rationales proposed are to:

- increase the likelihood of achieving objectives
- encourage proactive management
- be aware of the need to identify and treat risk throughout the organisation
- improve the identification of opportunities and threats
- comply with relevant legal, regulatory requirements and international norms
- improve mandatory and voluntary reporting
- improve stakeholder confidence and trust
- establish a reliable basis for decision-making and planning
- improve controls
- effectively allocate and use resources for risk treatment
- improve operational effectiveness and efficiency
- enhance health and safety performance, as well as environmental protection
- improve loss prevention and incident management
- improve organisational learning

The main issues related to risk management interventions are often concentrated around Small to Medium Enterprises because of limited financial resources, lack of appropriate competences and management tools and a burden of compliance with regulations and codes (Hasle & Sørensen 2011). However, SMEs are not the only type of organisations that could improve their performance by developing a capability to assess risks in actual operations and to monitor daily risks (Leva et al 2017). As highlighted in other chapters of the book, there are many methods that provide convincing arguments for the benefits delivered to an overall company performance. Examples include: the methods and examples provided for assessing risk in actual operations and the projects using dynamic event trees to test different procedures for carrying out safety critical activities (see Chapters 3 and 9), or the capability to continuously monitor risk performance to deliver better decision-making (see Chapters 6 and 7).

In this context, the use of decision analytics may be particularly crucial in the area of risk based decision-making. Wijnia (2016) proposes some important questions when applying decision analytics and cost benefit analysis (CBA) in the context of an asset management in a safety critical industry. The questions that Wijnia proposed are:

1 What is an adequate representation of the value system that facilitates CBA as the selection of most important risks?
2 What is an adequate representation of the risk position?
3 What is the effectiveness of applying Risk Based Optimisation by means of these adequate representations?
4 How robust is this effectiveness of Risk Based Optimisation?

In addressing the above questions, Wijnia performed an evaluation of several ways to include more risk based considerations into decision-making. Many of them could lead to the conclusion that a drastic change of the decision could be reached by using a risk based approach and in terms of the Total Cost of Ownership (TCO). The improvement delivered by risk based approaches was about 20% compared to the conventional approach. Furthermore, the study showed that a very robust decision could be achieved even without fully understanding the precise failure mechanism behind the risk, which means that the approach is not just for specialists.

Conclusions

A possible way forward for implementing a TSM approach would be to adjust risk management to add explicitly to the perspective of the CEOs. This means that the safety risks should likely always be mapped into the overall landscape of the organisation's goals and business strategies.

As pointed out previously, decision makers in companies may consider safety as a separate activity from their core business. One example is the identification of Key Performance Indicators (KPIs) separately from Process Safety Indicators (PSIs). A Cost Benefit Analysis (CBA) showing potential losses avoided (e.g. through reduced value-at-risk of the assets) and benefits obtained (e.g. more reliable revenues, maintained reputation, etc.) as part of a properly selected risk assessment approach can help top management to look at risk assessment and risk management more favourably. In other words, the idea is that CBA shall be regularly used in conjunction with risk assessment when evaluating risks, e.g., various risk mitigation options are to be compared by the management.[3]

Although companies have an obligation from authorities to perform risk assessment and to achieve some minimal safety criteria, this approach can highlight even further how an analysis of benefits could help companies to utilise their resources efficiently and safely.

By adding cost benefit analysis to the risk management process in a company's main business, safety would become an integral part of a company. To keep safety as an integral part of core business throughout an organisation, companies have to consider standardisation of their process for risk assessment and to streamline them into their option evaluation practice when performing CBA.

Safety is often considered a long term investment unable to provide immediate benefits. Therefore, the application of risk management tools needs to demonstrate the ability to generate business value event at the level of everyday operations.

The issue to tackle here is to convey to companies how they can improve their productivity tomorrow by investing in safety today. This is why the present book concentrated on the concept of Total Safety Management in order to shift the concept of safety from simply assessing risks or avoiding accidents into improving the effectiveness of operations.

In fact, in risk assessment and management studies, a new concept can also be introduced that can assess the available resources of organisations and their effective use in a best possible way, hence optimising industrial operations while providing for safety at the same level. In this way, safety assessment and its standardisation can help in achieving additional benefits while ensuring the safety of plant operations. These concepts need to be embedded in a cost benefit analysis of safety measures, as shown in the case studies presented in the present book regarding:

1 the need to support a common operational picture for the understanding of hazards and risks connected to business goals

(Chapter 2)

2 the benefits of standardising a process for risk assessment

(Chapters 3 and 4)

3 the capacity to understand/anticipate organisational trade-offs in safety management

(Chapter 4 and 5)

4 the benefits of developing a risk register to deliver risk intelligence (Chapter 6)

5 the capability to deliver total project planning by integrating task analysis, safety analysis and optimisation techniques (Chapter 9)

6 the benefits delivered by the capability to continuously monitor safety performance (Chapter 7)

7 the benefits of safety by design in practice (Chapter 8)

8 the aggregation approach on the KPIs reflecting safety performance (Chapter 9)

9 the advantages of knowledge management in total safety for major hazards plants (Chapter 10)

10 the effective new approaches for training in safety critical activities in the industry (Chapter 11)

11 the advantages of adopting a risk based approach in the management of technical and organisational changes in safety critical industries. (Chapter 12)

The authors hope the methods and the case studies presented are able to offer the necessary evidence and support to foster better practices for total safety management and risk based approaches at management level in companies and institutions.

Notes

1 In our experience such an attitude usually results in a safety case (safety program) being only a "dead letter on a paper" (forgotten) and attention is paid only when the authorities are coming for a visit or licensing procedure.

2 Please refer also to discussion in Chapter 1, the section "Business case for safety".
3 CBA should address also inherent uncertainties and unavoidable ethical issues in monetisation of the damages, see for example UK HSE's Guidance on ALARP principles at www.hse.gov.uk/risk/theory/alarpglance.htm .

References

Abad, J, Lafuente, E & Vilajosana, J 2013, 'An assessment of the OHSAS 18001 certification process: Objective drivers and consequences on safety performance and labour productivity', *Safety Science*, vol. 60, pp. 47–56. doi:10.1016/j.ssci.2013.06.011

Argilés-Bosch, JM et al. 2014, 'Empirical analysis of the incidence of accidents in the workplace on firms' financial performance', *Safety Science*, vol. 70, pp. 123–32. doi:10.1016/j.ssci.2014.05.012

Ayessaki, W-Y & Smallwood, J 2017, 'Influencing workers' performance through health and safety interventions', *Procedia Engineering*, vol. 182, pp. 42–9. doi:10.1016/j.proeng.2017.03.111

Brigham, A & Linssen, S 2010, 'Your brand reputational value is irreplaceable: Protect it!', *Forbes*, www.forbes.com/2010/02/01/brand-reputation-value-leadership-managing-ethisphere.html

Brousseau, KR et al. 2006, 'The seasoned executive's decision-making style', *Harvard Business Review*, vol. 84, no. 7–8, pp. 183–4.

Daft, RL 2008, *Management*, South-Western Cengage Learning, Boston.

Forteza, FJ, Carretero-Gómez, JM & Sesé, A 2017, 'Occupational risks, accidents on sites and economic performance of construction firms', *Safety Science*, vol. 94, pp. 61–76. doi:10.1016/j.ssci.2017.01.003

Hasle, P & Sørensen, OH 2011, 'When health and safety interventions meet real-life challenges', *Policy and Practice in Health and Safety*, vol. 9, no. 1, pp. 3–16.

Hopkins, A 1999, 'For whom does safety pay? The case of major accidents', *Safety Science*, vol. 32, no. 2–3, pp. 143–53. doi:10.1016/S0925-7535(99)00017-X

'How to live with risks' 2015, *Harvard Business Review*, vol. 93, no. 7/8, pp. 20–2.

ISO 2009, ISO Standard 31000:2009(en), Risk management: Principles and guidelines, viewed 15 January 2016, www.iso.org/obp/ui/#iso:std:iso:31000:ed-1:v1:en

Kaplan, RS & Norton, DP 2007, 'Using the balanced scorecard', *Harvard Business Review*, vol. 85, no. 7/8, pp. 150–61. doi:10.1016/S0840-4704(10)60668-0

Kim, BK et al. 2012, 'Case study analysis of the financial impact of catastrophic safety events', *Journal of Loss Prevention in the Process Industries*, vol. 25, no. 5, pp. 780–7. doi:10.1016/j.jlp.2012.05.006

Lafley, AG 2009, 'What only the CEO can do', *Harvard Business Review*, May, pp. 54–63.

Leva, M.C., Balfe, N., McAleer, B. and Rocke, M., 2017. Risk Registers: Structuring data collection to develop risk intelligence. *Safety science*, *100*, pp. 142–56.

Makino, R 2016, 'Stock market responses to chemical accidents in Japan: An event study', *Journal of Loss Prevention in the Process Industries*, vol. 44, pp. 453–8. doi:10.1016/j.jlp.2016.10.019

Mearns, K & Håvold, JI 2003, 'Occupational health and safety and the balanced scorecard', *The TQM Magazine*, vol. 15, no. 6, pp. 408–23. doi:10.1108/09544780310502741

Reason, J 1998, 'Achieving a safe culture: Theory and practice', *Work & Stress*, vol. 12, no. 3, pp. 293–306. doi:10.1080/02678379808256868

Rosenzweig, P 2013, 'What makes strategic decisions different', *Harvard Business Review*, November, pp. 88–94.

Soll, JB, Milkman, KL & Payne, JW 2015, 'From "economic man" to behavioral economics', *Harvard Business Review*, vol. 93, no. 3, pp. 65–71.

Spedding, L & Rose, A 2007, *Business risk management handbook a sustainable approach*, 1st edn., CIMA Publishing, Amsterdam, Boston.

Tappura, S et al. 2015, 'A management accounting perspective on safety', *Safety Science*, vol. 71, no. PB, pp. 151–9. doi:10.1016/j.ssci.2014.01.011

Veltri, A et al. 2013, 'Understanding safety in the context of business operations: An exploratory study using case studies', *Safety Science*, 55, pp. 119–34. doi:10.1016/j.ssci.2012.12.008

Wheelen, TL & Hunger, JD 2011, *Strategic management and business policy: Toward global sustainability*, 13th edn., Prentice Hall, Boston.

Wijnia, YC 2016, 'Processing risk in asset management: Exploring the boundaries of risk based optimization under uncertainty for an energy infrastructure asset manager', PhD Thesis, TU Delft, NL. doi:10.4233/uuid:e7fcf310-749f-4e4e-bbab-78eef99d0342

Yuliansyah, Y, Gurd, B & Mohamed, N 2017, 'The significant of business strategy in improving organizational performance', *Humanomics*, vol. 33, no. 1, pp. 56–74. doi:10.1108/H-06-2016-0049

Index

Note: Page numbers in *italics* indicate figures and page numbers in **bold** indicate tables.